Now What? A Call to Action

Environmental Systems Lessons for Upcoming Generations

Alan Ticotsky
in collaboration with the
Creative Learning Exchange

Alan Ticotsky and the Creative Learning Exchange
Acton, Massachusetts
2022

To my grandchildren, Caleb, Linny, Gabby, and the player to be named later, with a world of love

Cover photo by
Hand holding Earth (foreground) by Gerd Altmann
Trees and Clouds (background) by Adrian Pelletier

Funding was made possible through
the generosity of the
Gordon Brown Fund.

ISBN: 978-0-9960128-6-7 (Print Edition)
ISBN: 978-0-9960128-7-4 (Kindle Edition)

Print Edition ©2022
Printed through KDP
Digital Edition ©2022

Table of Contents

Acknowledgments

Writing *Now What?* has been a long journey and I could not have done it alone. Thanks to Lees Stuntz and Anne LaVigne, who spent many hours with me creating a book from a manuscript in need of lots of work. Their patience, wisdom, and persistence deserve my utmost gratitude. Lees has provided financial support, intellectual growth, and warm friendship. Her leadership continues to help steer the systems education movement. Anne is a treasure within the K-12 systems community.

My long time colleague and collaborator Rob Quaden has continued to be an inspiration to me in teaching and learning. Rob is one of the finest educators I have been privileged to work with and a valued friend.

As Head of School, Greg Orpen offered me a unique opportunity to work with students and teachers at Innovation Academy Charter School, an amazing public middle and high school where the values of systems citizenship are practiced and expected. Greg is among the most important systems thinking educational leaders in the world.

The teachers and students at Innovation Academy welcomed me to their classrooms and helped inspire this book. While I cannot name everyone there, a special shoutout goes to the high school science department for their kindness and collaboration, especially Chris DiCarlo, Katharine Hinkle, Raksmey Derival, and Ashley Joyce.

Thanks to Davida Fox-Melanson. As my supervisor and friend, she encouraged me to stretch and become a writer and mentor as well as maintaining the values to do good and to do well as a teacher.

The generous support of the Waters Foundation (now the Waters Center for Systems Thinking) allowed me to learn and grow in the fields of systems thinking and system dynamics. I am grateful for the trust shown and the opportunities provided during my eight years as a mentor.

Many system dynamicists have contributed their time and expertise to those of us in the K-12 education community. Their support through workshops, student expositions, conferences, consultations, and other modes of collaboration is essential to our work in general and to my understanding complex ideas.

I thank my family for their love. A lot of the work on *Now What?* happened during the COVID-19 pandemic, and my wife Jane and I spent almost all our time together. She continues to prove that I am a lucky man to be with her every day. Our children and their spouses also brighten my life, and as the dedication indicates, our grandchildren have helped keep me focused on "*Now What?*"

Introduction

"I realized that the most interesting problems in the world –energy, population, environment, food, international security – lived in the gulf between the social and physical sciences." John Holdren

Purpose and Goals

Environmental crises and world systems interconnect.

The title *Now What? - A Call to Action* draws us into one of the most important questions of our time. The world faces a decision point in the 21st Century. Major "Environmental Systems" (FIGURE 1) are becoming so stressed that our future quality of life may decline after hundreds of years of growth and development. But strategies and choices exist that lead to a happier outcome. In this book, environmental systems lessons contain activities and research about sustainability, land, air, and water for students to explore to create an environment that has the ability to regenerate as well as sustain.

Because "the environment" is not a single, isolated system, all aspects of civilization are bound together in an interconnected web. Social justice, public health, and economic prosperity all depend on a healthy environment that is consistently regenerating.

Societies must address "Social Justice" to ensure peace and prosperity for all people. Pandemics and other "Public Health" threats inflict terrible damage and disrupt life around the world. Problems providing "Economic Prosperity" exacerbate income inequality and degrade the quality of life for many families. What is to be done? While studying current topics through a systems thinking lens, students are encouraged to make connections between environmental systems and other social systems and to research strategies for positive change.

FIGURE 1

Educators have leverage.

As educators we can use our leverage to motivate younger generations. Promoting and protecting a healthy, regenerative environment is not easy and changes must be implemented.

Younger generations need to be reassured that positive change can lead to a sustainable and prosperous future using the power of healthy systems to regenerate our environment. The situation isn't hopeless. While the interconnectedness of the world's problems creates great complexity, progress in any area affects the other systems as well.

The power of systems thinking
The lessons and activities in *Now What? A Call to Action* focus on environmental systems. Students use systems thinking tools and principles to analyze problems and make informed decisions about strategies and policies for the future. They are encouraged to think deeply about fundamental solutions and find leverage points to move toward sustainability, justice, and prosperity.

Pedagogical Philosophy

Active engagement

- Project-based learning
- Hands-on activities
- Simulations
- Role-playing
- Diagram and model building

In these lessons, **learning by doing** is emphasized. Activities are designed for students in high school and middle grades natural science, social science, and humanities classes. Students tackle actual contemporary problems, often playing roles such as consultants, diplomats, government officials, concerned citizens, company executives, and other people in simulated, real-world situations. They are asked to make persuasive presentations based on factual information, using systems thinking diagrams and other graphic tools to illustrate their positions.

Teamwork, problem solving, and interdisciplinary learning
Today's students face increasingly complex topics, in environmental studies as well as other learning contexts. They will be expected to make informed and difficult decisions as adults. In the lessons in this book, they work collaboratively as active learners and problem-solvers, just as they will need to in the "real world," becoming members of organizations that learn together.

Encourage revision and teamwork

People in almost all fields commonly revise their work numerous times before considering a piece ready for display or submission. Systems thinking diagrams are no different, so encourage students to review and revise their systems diagrams, just as they do with the work they produce in other disciplines. Almost all learning is an iterative process, so drafts of causal loops, stock/flow diagrams, and models will generally improve with further reflection and input. Systems diagrams represent mental models, and the more experience and feedback we have, the clearer those models become.

Systems educators find that students usually build the most useful and accurate models and diagrams when they collaborate in teams. As students share their work, feedback given should refer to the diagrams and to the facts and ideas that underlie them. Discussion that focuses on specific diagrams helps make classroom conversations collaborative and encouraging, avoiding confrontational discourse. Civil behavior leads to better understanding among groups and creates pathways to consensus and problem solving, a goal for our lives in schools and elsewhere.

Learning about environmental systems covers content associated with many subjects, ideal for an interdisciplinary approach:

- Physical Science – Earth science, chemistry, physics, biological sciences
- Social Science – history, politics, current events, social justice, economics, public policy, geography
- Mathematics – quantitative skills, computer modeling

- Humanities – language skills, reading, writing
- Technology and Graphic Design – presentations, computer modeling
- Health – effects of environment, nutrition, medical science

Using systems thinking to broaden our understanding of a sustainable world in which the regenerative capacity of our ecosystems is restored.
The lessons in this book employ systems tools and principles as integral parts of interactive and engaging activities. Systems thinking tools guide students throughout the learning process:

- choosing and framing essential questions
- collecting and analyzing information
- investigating change over time, causality, feedback, delays, and other characteristics of complex systems
- producing effective presentations with clear graphic representation of ideas

"I find that I read the newspaper differently."
After people study the principles and tools of systems thinking, most comment that they read or hear news differently. They use less linear, cause→ effect thinking when approaching complex problems. Instead, they look at dynamic behavior, change over time that has brought about conditions that need to be addressed. Finding feedback loops within the structures of systems highlights leverage points, places where policy change can make a difference.

"Trust the students"
They learn fast and are often eager to jump in and try something new. Eventually, the tools become part of their repertoire.

How to use this book

The Lessons
Each lesson and each chapter can stand alone or be taught in units as part of a comprehensive environmental education curriculum. Each chapter takes on a different aspect of environmental systems. Chapters begin with lesson previews and conclude with debriefing guides. All lessons focus on systemic problems. Some are historical, some hypothetical, and one is fictional. Many involve "current events," problems this generation must confront. Suggested articles are potential starting points for research, and students are encouraged to investigate further independently.

Learner-centered Learning

It is important to allow source materials to speak for themselves. Teachers should activate students' prior knowledge and curiosity when beginning to study a topic so students can see if their opinions change. The quotations at the beginning of each lesson will serve to activate student thinking. Encourage questions and have students follow their interests when researching. Experts (and non-experts) look at climate change, for example, from many viewpoints; scientific, political, economic, social, historical, and geographical, just to name a few. Students should bring their own thinking to any issue and then direct their learning to historical trends, current events, and future projections.

The Appendices
Information and handouts in the appendices contain useful material for both students and teachers.

Appendix A: Connections illustrates ways that the lesson content relates to current educational standards. It's also a handy guide to see where different tools, principles, and topics come into play.

Appendix B: Lesson Resources contains links to suggested articles and other needed resources to complete the lesson activities.

Appendix C: Systems Resources includes introductory materials for systems thinking tools and handouts in the form of templates. Students can of course draw out any of the tools on blank paper, but it can be useful to use the templates when first learning the logistics of how to use any particular tool.

Appendix D: Instructional Resources contains additional suggestions for student projects and assessment, including potential checklist-style rubrics.

Tools and Resources

"Current events"

Students take on environmental systems problems in most of the lessons. The materials listed within these lessons include suggested articles or websites, which are linked in Appendix B. Many of the resources contain video and audio media. The New York Times and other commercial publishers are cited frequently, as well as United States government websites and additional sources. Guide students to be mindful of bias within websites and urge them to use multiple sources by entering key terms from the article titles or summaries into a search engine.

Student research - finding articles and reputable sources

In addition to the suggested articles/topics, students are encouraged to do their own research. Be sure they choose articles from credible sources that are known to check their facts.

Many articles from commercial publishers are protected by a paywall. After a certain number of articles read per month, non-subscribers aren't allowed to download and read any more. Digital subscriptions are often much less expensive than paper copy subscriptions but still may be too costly for teacher or student budgets.

Debriefing the Lessons

Every lesson and chapter concludes with debriefing questions and summaries. Generally, the questions are open ended enough so that the phrase "Answers will vary" applies universally. The questions encourage discussion and provide students with opportunities to organize and present their research and thinking. The answers given in italics serve as a general guide for teachers and employ systems thinking perspectives. Encourage students to connect current learning to experiences and prior knowledge from previously completed lessons and/or chapters.

Consider asking students to reflect on generic, big picture connections and systems thinking principles after each lesson, perhaps in a journal or in group dialog sessions. Questions might include:

- How is this activity/lesson similar or different from previous lessons experienced so far this school year?
- How does the opening quotation relate to the main theme of the lesson?
- What patterns or themes can we identify?
- What connections can we see between this activity and previous ones?
- Given what we just learned/experienced, what causal connections and feedback loops exist among sustainability, air, land, and water systems?
- What new understandings and insights have we developed, given our cumulative studies of environmental systems thus far?
- What systems thinking skills have we been accessing or should we be accessing while working to improve environmental systems in our local community? Why are these important?
- What questions do you have after studying this topic?

Suggestions for accessing news articles and features:

- Check school and local libraries for subscription privileges.
- Have students in a group download different articles, stretching their monthly allotment.
- Put an article title or topic summary into a search engine and choose among results. Articles may be reposted and available free of charge.
- Investigate other media besides periodicals, like blogs and video talks.
- Don't forget printed materials – books, magazines, newspapers, etc.

Using systems principles and tools

We all struggle to understand complex problems. Systems thinking principles, skills, and tools provide methods and frameworks for analyzing them, seeking their causes and the strategies leading to solutions.

Here is a list of some important principles of systems thinking:

- Everything is connected – there is no "away."
- Everyone uses mental models to make decisions.
- Nothing grows forever – systems tend toward equilibrium.
- Systems are usually resilient, resisting change but may still be pushed out of equilibrium.
- Feedback is embedded in all actions.
- Cause and effect are often not closely related in time or space.
- There are seldom simple solutions.
- Solutions often become problems.
- Actions may be ineffective due to low leverage policies or incorrect application of high leverage ones.
- What appears to be good in the short term is often bad in the long term, and vice versa.
- Almost nothing is exogenous – causes of problems are within systems (endogenous).
- Systems and problems within them may display parallel structures even though the contexts are unalike (generic structures).

The Systems Thinking Skills of Barry Richmond are a major unifying theme throughout the book. Lessons 1.1 and 1.2 introduce the skills and they remain an integral component in subsequent lessons. Barry devoted much of his life to helping develop people's capacity to be "systems citizens." Systems citizens consider themselves "members of a global community. They strive to understand the complexities of today's worldly systems and have the ability to tackle problems with an informed capacity to make a positive difference" (from *Tracing Connections*).

Barry Richmond's Systems Thinking Skills

10,000 Meter / Forest Thinking – Seeing big picture rather than focusing on details

Dynamic Thinking – Recognizing patterns rather than events, realizing that continuous processes rather than discrete events create the conditions in the real world – understanding that you can always quantify, though you can't always measure

System as Cause Thinking – Placing responsibility for a behavior on internal actors, self-generating rather than external forces

Operational Thinking – Understanding how things really work, e.g., maintaining consistent units of measure, and expecting that factors don't always cause proportional effect

Closed-Loop Thinking – Tracing ongoing interdependent processes rather than one-way causal chains

Scientific Thinking – Testing hypotheses, applying rigorous simulation and quantification

Empathic Thinking – Giving and receiving feedback, expanding personal boundaries, avoiding taking a single rigid position and appreciating a spectrum of opinions

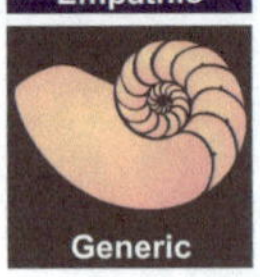

Generic Thinking – Recognizing generic structures rather than specific events

Systems Thinking Tools

Systems thinking promotes "big picture" perspective while remaining grounded in practical, operational strategies for understanding complex problems. Systems thinking analysis focuses on change over time and feedback relationships among the elements of systems, so students will be learning to consider causality in broader, non-linear terms compared to traditional cause and effect thinking. Ideally, students will become adept at using a variety of tools and apply each one where it's useful and appropriate. But all the tools move thinking in a deeper direction and lead us to ask good questions.

Each tool is like a different door into the world of systems thinking. Support for using the tools is provided in the appendix, through instructions and sidebars in the lesson plans, and via links to helpful sites.

All the tools describe the same behaviors and situations in slightly different forms.

Behavior over time graphs (BOTGs) are a good starting point and should almost always be part of the analysis of a topic or problem. Students should develop the habit of graphing variables over time, using whatever data are available or what assumptions they are making. BOTGs require little technical skill and provide a lot of information.

Students will be able to ask the four basic questions for BOTGs in every lesson:

- What is changing? Identify the variables.
- How is it changing? Describe the direction and rate of changes.
- Why is it changing? Investigate causality and feedback.
- So What? and Now What? That is, what is important and needs to be done?

Connection circles are used to discover causal connections in a system and to find feedback loops. Connection circles usually serve as a precursor to causal loop diagrams.

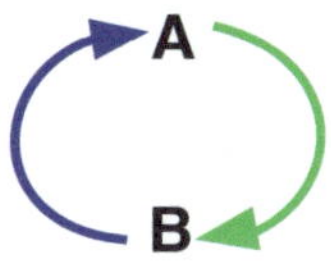

Causal loop diagrams (CLDs) display feedback loops in a system. CLDs are either reinforcing, meaning the behavior continues in the same direction, or balancing, in which the behavior oscillates, or changes between increasing and decreasing over time.

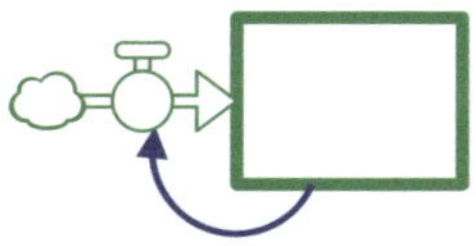

Stock/flow diagrams (S/F) represent accumulations and rates of change of "stuff" in a system. S/F diagrams help us see how a system actually operates and changes.

System dynamics computer models enable us to simulate behavior in a system. Once equations are entered, models may be run to ask "what if" kinds of questions.

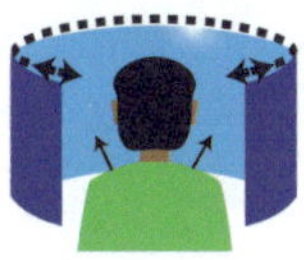

Simulations help compress time and space, so rather than having to wait a hundred years to see a particular result, a simulator can reveal that result quickly and in a contained space, either real or virtual.

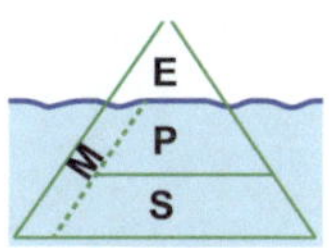

Iceberg diagrams graphically display how events may be seen in the context of longer-term patterns of behavior, and how those behaviors result from structural factors in the system. While events may be most noticeable and visible like the tip of an iceberg above the waterline, the deeper layers provide more insight and leverage for change. Influencing each layer are mental models, the views we all construct of the world.

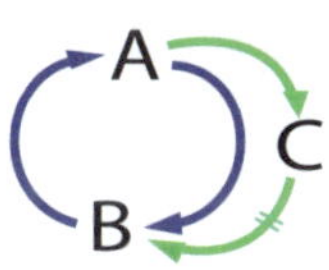

Systems archetypes (and generic models) are common patterns of behavior that are seen across disciplines in a variety of systems. Systems archetypes are usually drawn as causal loop diagrams.

Each kind of diagram, except for the systems archetypes, appears first in Chapter 1 and then throughout the rest of the book. Systems archetypes are introduced in Chapter 2. Each chapter contains a computer modeling activity or a simulation based on a computer model.

Software

Lessons often include drawing system dynamics diagrams. See Appendix C for information about system dynamics software options. All the diagrams in this book were created using Stella software by isee systems, inc.

Students may draw diagrams by hand, but using software allows them to save and share work easily and will look "professional" in their presentations. Building working models that run simulations can elevate their thinking and allows for testing "what if …" scenarios.

Support and Sharing

Help build a network

The community of systems educators is growing. Consider subscribing to the listserv on the Creative Learning Exchange website (http://www.clexchange.org/cle/k12listserve.aspx) and join a network of educators with common interests and purpose:

- learning, teaching, and practicing environmental sustainability
- integrating systems thinking and system dynamics into school curricula
- helping students learn how we got where we are and what needs to be done
- encouraging systems citizenship

You will receive posts and be encouraged to contribute content.

Updates to Appendix B will be available on the Creative Learning Exchange website at http://www.clexchange.org/NowWhat/, with additional links to curriculum and other resources.

Good luck and remember to trust the students. They are our future.

Chapter One: Sustainability, Systems Thinking, and the Environment

"People who embrace systems thinking position themselves such that they can see both the forest and the trees."
Barry Richmond

Getting started – Environmental studies, sustainability, and systems thinking

Environmental systems are complex and the world faces important problems without easy solutions. Sustaining a healthy and prosperous environment can only happen when citizens think systemically and accept responsibility. Finding common ground among varied points of view may sometimes seem impossible, but often people who disagree about strategy actually seek similar goals. Conflict occurs frequently because of short-term policy differences and the disparity of resource allocation. Taking a long-term view and thinking systemically helps us understand the position of others and look for lasting solutions to difficult problems. Barry Richmond called this thinking process "becoming a systems citizen."

What will students do and learn in this chapter?

1.1 Sustainable Living in Cities
Consider cities through the framework of systems thinking skills developed by Barry Richmond.

1.2 Transportation – Moving Efficiently
Investigate transportation issues in urban areas using the systems thinking skills.

1.3 Electrical Energy
Research how people use energy and reflect on potential pathways moving forward. Systems tools to use include behavior over time graphs, connection circles, and causal loop diagrams.

1.4 Recycling
Follow the path of solid waste, diagramming with stock/flow diagrams. Research methods for more sustainable practices.

1.5 Environmental History
Learn about crises and turning points in the 20th and 21st Centuries. Study graphs, and choose among feedback loops, iceberg diagrams, and stock/flow diagrams to help explain what happened and find pathways to avoid future damage.

1.6 Modeling Population
Gain a basic understanding of population dynamics using simulation activities, graphs, stock/flow diagrams, and computer models.

1.1 Sustainable Living in Cities Using Systems Thinking Skills

Environmental Systems

- How can urban areas be managed sustainably?

Tool

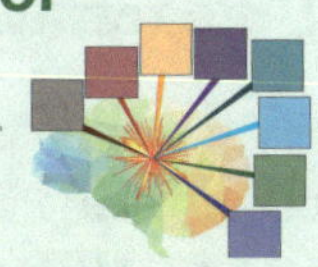

Skill

"The city is the nerve center of our civilization." Josiah Strong

Overview

Urban areas continue to grow and house an increasing percentage of the world's population. How can we use the advantages of cities to build a sustainable future?

Students learn about Barry Richmond's Systems Thinking Skills. They use the skills to consider policies for sustainable urban development, presenting ideas while participating in a simulated urban planning roundtable.

Materials

- Suggested articles, p 198
- Handout: Attributes of Cities, p 17
- Resource: Systems Thinking Skills, p 214

Procedure

1. Ask students to imagine they are being hired to help plan future development in a large city. Have them begin by brainstorming their ideas about the strengths and weaknesses of cities, working in teams or pairs. To help them surface their mental models, provide the included handout or suggest they make a chart in the shape of a 'T' and list some positive and negative attributes like the examples in the table.

Barry Richmond and Systems Thinking Skills

Barry Richmond (1946 – 2002) was a leading figure in systems thinking and system dynamics. During his professional life, he taught higher education courses, formed a pioneering company that developed modeling software and provided training and consulting, wrote eloquently about systems, and became a valuable and generous friend to K-12 educators. Over time his writings described several versions of essential systems thinking skills. In this activity, students will use eight skills Barry described. Please see the resources section for links and reference to Barry's important and inspiring work.

See Appendix C for more detailed explanations of these system dynamics tools and links to further learning.

Positive attributes of cities	Negative attributes of cities
Rich cultural offerings	Crowded and noisy
Transportation hub	Lots of traffic
Many available services	Limited open space
Large workforce	Many sources of pollution

2. Assign articles about sustainable cities to students or encourage them to find their own sources. In pairs or small groups, have them summarize their article for their partners. Then they should add any major topics related to sustainability that weren't in their T diagrams.

3. After sharing and saving the lists of positive and negative attributes of cities, distribute Barry Richmond's Systems Thinking Skills, which is at the end of this lesson plan. Each skill is explained and contrasted with a more narrow way of thinking. The systems thinking skills emphasize a broad view rather than focusing on one aspect of a system.

   Challenge each team to give examples of how one or more systems thinking skills can relate to the problem of sustainable cities. How can they use the skills to advocate for specific city planning initiatives of their choice? Emphasize that the skills don't provide answers to problems but rather ways to frame our thinking about complex issues.

4. Use the debriefing questions in the Bringing the Lesson Home section to guide the teams of student consultants to present their reports telling how they would help fashion an effective and attractive sustainable city. The sharing can take a variety of forms – posters, slide projection, oral presentations, written reports, panel presentation, or other format. Conclude with a whole class planning roundtable, surfacing the most promising ideas from each team.

   The following pages have sample ideas for applying each thinking skill to urban planning topics. Use these examples if any team gets stuck, or share one to provide a starting point for students' own thinking.

Feedback

Most people have experience with types of feedback – for example, commentary given about performance or ideas, or unwanted sound being amplified by a microphone. Systems thinking deals with changes over time and the causal elements that drive those changes, and feedback has a special but similar meaning for systems thinkers.

Feedback operates as a loop mechanism. An action or input causes a result, which is itself another action. That result in turn causes other actions and eventually the original input will be affected as part of a closed loop. Systems thinking emphasizes the importance of recognizing and understanding feedback, a skill Barry Richmond called "closed loop thinking."

Examples of using Barry Richmond's systems thinking skills: generating questions and suggestions for urban planning

Seeing the big picture rather than focusing on details

Seeing the Earth from an airplane 10,000 meters above the ground allows one to see more of the forest, not just the individual trees. Urban problems need systemic solutions:

- Too many cars – building parking lots may ease congestion in one neighborhood, but probably draws in more drivers to the city. The big picture demands broad ranging solutions, including public transit, staggered working hours, working from home, and alternatives to cars.
- Promoting healthy lifestyles – access to fresh food, clean air, open space for exercise and relaxation, affordable health care, and a goal of overall well-being.

Tracing ongoing interdependent processes rather than one-way causal chains

Sometimes we are conditioned to think in strictly linear terms:

But usually that effect becomes a cause for another effect, and eventually the first cause in the chain is itself affected.

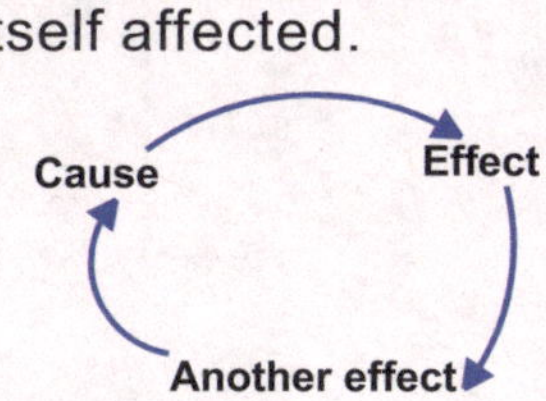

In the example about streets in operational thinking on the next page, consider "Concern about streets" as a cause. The effect is a "desire for improvement" and then a series of 'events-becoming-causes' creates a closed loop. When the quality of streets improves, it can have the effect of making "concern about streets" decrease. Closed-loop thinking helps us recognize feedback in systems.

Recognizing patterns rather than events, realizing that continuous processes rather than discrete events create the conditions in the real world – understanding that you can always quantify, though you can't always measure

- "This week there were traffic jams almost every morning!" That's an event in a short time frame, but it may be indicative of patterns and trends. Can you cite data that report increases in travel time and number of vehicles on the road over time? Is traffic a problem during one specific time or across the day?
- Population, economic data, movement of people, weather, and climate ... Lots of information can be gathered about cities. Other important aspects of urban life can't be measured with tangible units. To assess happiness or satisfaction, a researcher may ask respondents to use a scale of 1 to 10, for example, and graph the results. We all have mental models that need to be explained and tested at times, and dynamic thinking can be applied both to things we can count and those we cannot.

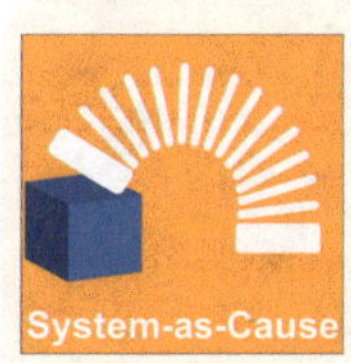

Placing responsibility for a behavior on internal actors, self-generating rather than external forces

Systems thinking teaches that "structure drives behavior." Instead of blaming people with bad intentions for problems, look for causes within the system.

- Housing costs are high in many cities because a lot of people are looking for places to live.
- Some cities are troubled by litter because trash containers are designed poorly or aren't spaced correctly to encourage responsible disposal. Perhaps recycling programs are difficult to use.

Understanding how things really work, e.g., maintaining consistent units of measure, and expecting that factors don't always cause proportional effect

- Preserving green space for people in cities is important. However, looking at the percentage of open space or parks won't always tell the whole story of how effectively the parks are used. Parkland will not benefit the recreational needs of citizens if they cannot access it. How close to population centers are the green spaces? Can they be reached by public transportation? What facilities are available for people in all neighborhoods?
- While we might predict that "Policy A causes Effect B," things seldom happen exactly as planned in the real world. Let's say a study group predicts that raising tolls will reduce traffic congestion by a certain percentage. Their reasoning forecasts that fewer drivers will be willing to pay more, leading to car pools, increased public transit ridership, and other alternatives to automobile commuting. Operational thinking prepares us for a variety of outcomes, including:
  - ◊ Increases in tolls might not have proportional decreases in drivers.
  - ◊ After an initial drop in drivers, some people may be attracted to a less congested commute and decide the higher toll is worth paying.
  - ◊ Traffic may transfer to side streets, shifting congestion away from the toll roads.

- Structure drives behavior. In order to create an increase in something, units of that thing must accumulate. Let's say a city has a problem with the quality of streets: potholes, broken pavement, faded lane lines, and so on. Concerned citizens may conduct a campaign to improve the situation, beginning with publicity to gain support for change. Will more concern about streets improve their quality?

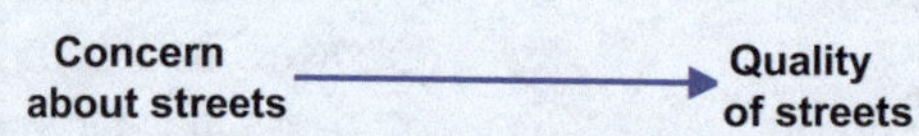

- While concern may grow as support increases for improving street quality, those "concern units" don't build better streets. Wishing won't make it happen. But when concern is great enough, desire for change grows. People are spurred to act, and money is allocated for street repairs and improvements. Work is done, and the streets improve. The resulting causal loop diagram draws a pathway from concern to action. All the steps in the causal loop are important, but work being done on streets is what actually improves them.

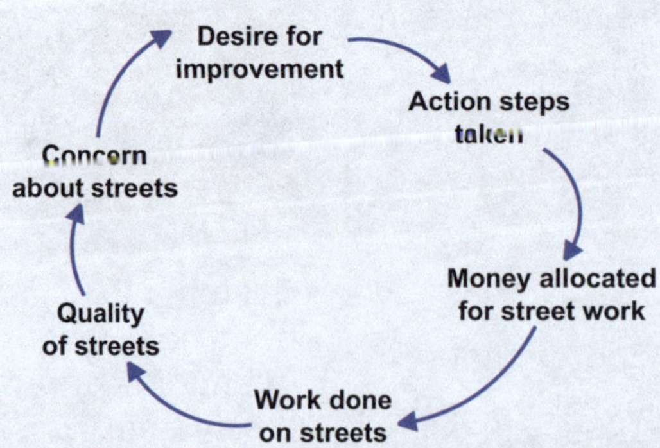

Note that the expanded diagram connects "Quality of streets" to "Concern about streets," creating a loop. Operational thinking helps with another thinking skill, closed-loop thinking.

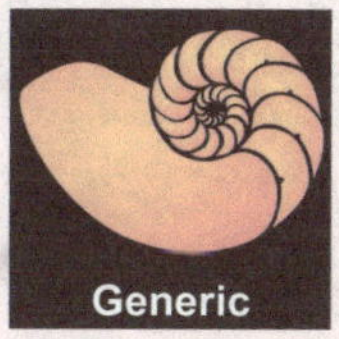

Recognizing generic structures rather than specific events

Systems thinkers look for ideas that cross disciplines. Economies of scale can create significant advantages for cities, just as they can for companies, individual animals, someone's garden, and many other examples. The cost of producing something per unit often goes down when resources are concentrated and more units are produced. The disadvantages that build up when an organization or other entity (like a city) grows too large must be accounted for also. Is there a 'sweet spot' for each city - large enough to be vibrant and efficient but not unmanageably crowded and complex?

Testing hypotheses, applying rigorous simulation and quantification

Modeling can be very useful before implementing policies. Officials and consultants need to research the effect of relevant factors and run simulations to predict the likely effects of changes. Using data is vital to evaluating existing situations and making decisions moving forward. What are the most pressing problems and where should resources be deployed to improve the quality of life?

Giving and receiving feedback. Expanding personal boundaries and appreciating a spectrum of opinions

Understanding others takes effort. Listening to their ideas and hearing their feedback is vital to solving complex problems. We often assume we understand the positions of others so it's important to surface everyone's "mental models," which Peter Senge describes as internal images everyone creates and holds about the world. City residents can feel very differently from commuters about how well aspects of a city reflect their needs and hopes. People with a range of personal wealth or those living in different neighborhoods can also experience their city in quite disparate ways.

Consider a situation in which a plot of land becomes available. How does a planning group decide how to use the land? Perhaps a farmers market can be set up to provide fresh produce for residents who become customers for growers. Maybe there are no supermarkets nearby, and a shopping area should be developed. Are there housing needs to be met in the neighborhood by building residential units on the land, or should the location be developed for recreation and open space?

Most issues cover a spectrum of opinions and positions. Complex problems seldom have simple solutions. Strategies must account for the diversity of factors by avoiding rigid thinking.

Bringing the Lesson Home

What is the key difference in the systems thinking skill(s) your team used compared to other ways of thinking about problems? How could the systems thinking skills help design a city?

Have each team expand on the short phrases written on their handout. Check on their understanding by asking for examples relating to sustainable cities. Each team will highlight aspects of planning that work best with their systems thinking skills. Look for common themes and attributes across the groups as the skills overlap. Attributes may include:

- *Long-term and short-term thinking and planning*
- *Considering internal causes with feedback*
- *Looking at the big picture*
- *Being realistic and scientific*
- *Seeking rigorous data*
- *Maintaining a flexible and inclusive mindset*

What strategies or initiatives would you propose for a sustainable city based on the ideas from your team and others?

Teams will have had different approaches. Gather the specific ideas that emerge and continue to revisit them as students study sustainability and environmental systems in subsequent activities.

Create categories of action cities need to undertake to become or remain sustainable. These include housing, transportation, environmental health, jobs and wealth, education, and many more. Aspects of a system are interconnected and must be considered as a whole in order to address complex problems.

Handout: Attributes of Cities

Positive attributes of cities	Negative attributes of cities

1.2 Transportation - Moving Efficiently
Applying Systems Thinking Skills

Environmental Systems
How can metropolitan areas maintain efficient transportation systems?

Tools

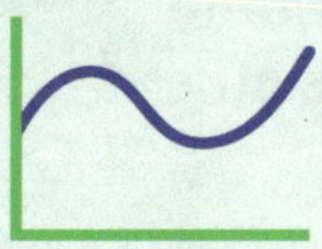

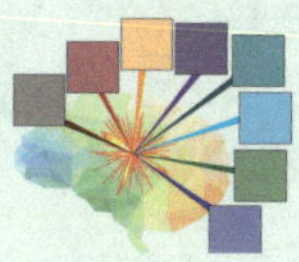

Skill

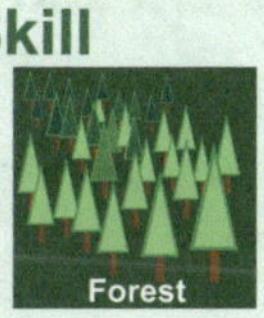

"A developed country isn't a place where the poor have cars. It's where the rich use public transportation." Gustavo Petro

Overview

Students take the roles of city transportation planners, focusing specifically on one aspect of urban life. How will they address the problem of congested roads, especially during rush hours? Using the techniques outlined by Barry Richmond's Systems Thinking Skills, students seek solutions that go beyond constructing more roads. In addition, can improvements be achieved if issues like transportation are planned without looking at overall sustainability?

Complex systems like urban transportation require people to think creatively. What constitutes a systems thinking mindset and viewpoint? Barry Richmond's set of skills presents a roadmap for understanding complex systems and encourages us to ask good questions.

Barry Richmond's "Systems Thinking Skills"

Lesson 1.1, "Sustainable Living in Cities," introduces students to many of the major principles of systems thinking described by Barry Richmond. The "Systems Thinking Skills" offer frameworks that will encourage students to use systems thinking to move toward creating a sustainable future.

Lesson 1.1 asks students to assume the role of city planners and provide guidance for a sustainable future. This lesson, 1.2, calls on them to focus specifically on a transportation issue, i.e., overcrowded roads causing traffic jams.

Materials

- Handout/Resource: Systems Thinking Skills, p 214
- Resource: Intro to BOTGs, p 215

Procedure

1. Organize students into pairs or teams. If students do not have copies from Lesson 1.1, distribute Barry Richmond's Systems Thinking Skills. If they did work on Lesson 1.1, tell them their good work as consultants has earned them another assignment. City officials and

citizen groups were particularly impressed with their suggestions and explanations using the thinking skills.

2. Present students with the problem of traffic congestion. It can be an actual description of local conditions students are familiar with, or use the following hypothetical situation:

   A few years ago, a two-year construction project widened a major four-lane highway between a city and the nearby suburbs. An additional travel lane was added in each direction. When work was completed, traffic moved more smoothly for a year or two. But gradually jam-ups and delays began to occur during rush hours again. Congestion became just as bad or even worse than before the expansion of the highway. Some individuals and groups are advocating for another construction project while others argue against it. What should be done?

   Have each team generate ideas and questions planners should consider, using one or more of the thinking skills as a guiding principle. As consultants, teams should prepare posters and/or slide presentations of their work to share.

3. The description of systems thinking skills at the end of this activity provides some suggested perspectives on the highway question. You can decide how much guidance to give teams.

4. Set up a whole class sharing of ideas. Teams can participate in a simulated transportation forum, with groups presenting their posters and describing ideas they think will relieve traffic congestion and improve the flow of travelers. Keep the systems thinking skills handout as part of a reference collection for future activities.

Examples of using Barry Richmond's systems thinking skills: generating questions and suggestions for transportation planning

How does the highway fit in the overall transportation system? Are there public transportation options? Look at the whole metropolitan area rather individual roads. Try to think beyond seeing the problem as one of inadequate roads.

Will improving roads actually alleviate traffic congestion? What other assumptions contribute to choosing a specific plan for solving this problem?

Ask four key questions about a dynamic (changing) system:

1. What is changing? Identify the variable.
2. How is it changing? Look for the pattern of change.
3. Why is it changing? Identify causes.
4. So what and now what? Consider what is significant and what should be done.

What has been the change of traffic use over time? What is the daily pattern? Study graphs of traffic and commuting times. What might happen in the future?

Are there smaller scale projects short of road construction that can make an incremental difference, even if a complete solution doesn't result? What smaller scale projects and initiatives might relieve congestion, avoiding the disruption of major construction? When we begin to make changes, are we on the right track?

Improvement is usually an iterative process. The phrase "successive approximation" is used in math and science for an incremental approach to problem solving and change.

Once we surface mental models held by ourselves and others, how can we evaluate whether our assumptions are reasonable and how they are influencing our behavior? Have things improved? Are our conceptions of causal relationships in line with what is actually happening? What are some available, easy to implement ideas to test to ease congestion?

How can we surface "mental models" to get a sense of what's actually happening? It's important for people to know where others "are coming from." What is a reasonable expectation about getting from one place to another and for driving a car? Can most commuters drive their own cars and travel within the same narrow time frame? Is it necessary to work the same hours every day, and be in the same place? What other assumptions about work and commuting underlie the structure that creates traffic congestion in metropolitan areas?

The elements of a system are connected and affect each other. A congested traffic situation requires that we look beyond the roads for ways to improve the flow of vehicles. What policies will bring desired changes? What leverage points will create the desired results?

What is changing in the system? How are the elements that increase and decrease affecting each other? System dynamicists refer to these elements as "stocks," and their rates of change as "flows." Stocks and flows describe the way a system actually works.

In our traffic example, the stock of cars on the road is related to population, economic growth, and other accumulations that change over time in a region. How do they affect each other? How can we develop policies that influence changes in flows that affect stocks?

What would be an effective motivation to entice people to share rides or use public transportation? What has been tried in the past? What ideas are likely to be successful and produce lasting improvement? Are we solving the "right" problem?

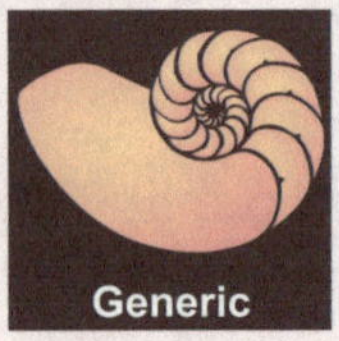

Systems thinking teaches us to recognize many characteristics of complex systems.

What is good in the short term is often not good in the long term, and vice versa. Short-term pain can bring long-term gain. Systems thinkers know things often get worse before they get better, and the opposite is also true. For the first few years, traffic may improve. But that may lead to increased usage, and soon we are facing the same problem or worse. What changes might be worth implementing for lasting improvement?

Actions meant to achieve solutions often have unintended consequences.

What are tradeoffs involved in taking action? Is the cure worse than the problem? Outcomes are usually difficult to predict, so it's imperative to weigh actions carefully. Building new bridges and roadway lanes can be very expensive and disruptive. Neighborhoods suffer when roads cut through them, bringing noise, air pollution, displacement of homes and businesses, and other negative impacts. Money designated for road construction could be spent elsewhere. What level of inconvenience is tolerable for a significant payoff later? Expanding roadways might undermine public transportation. People may choose to live farther from their jobs if they think commuting conditions have been improved. Many unintended consequences are not discovered until after a "solution" has been implemented, so careful consideration of outcomes is important albeit difficult.

Cause and effect are often separate in time and space.

Policies that advocate changes in behavior may not motivate people immediately. Delays are common in most systems. Cleaner air, reduced noise, and open space will attract people but it may take a while for them to change their behavior. Mass transit ridership might not grow in the first year or two after improvements are implemented, and the process can be very expensive. But an investment in trains and buses pays dividends well into the future, while road construction often becomes more of a problem than a solution.

Who are the different people who have a stake in the situation? How are their points-of-view similar and how do they differ? Does everyone drive to work? What else can be done to motivate people not to drive in their own cars?

Commuters travel paths between their homes and the places where they spend their day. Urban planners must consider housing and employment along with transportation in order to achieve the balance necessary to create and maintain a healthy and vibrant city. Transportation is part of urban planning.

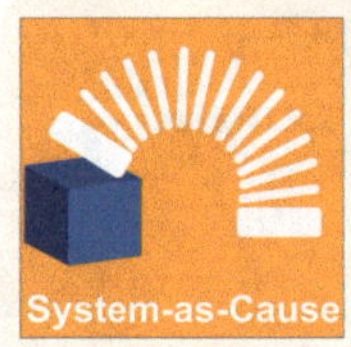

In most cases, structure drives behavior and behavior further supports the structure, creating a reinforcing loop.

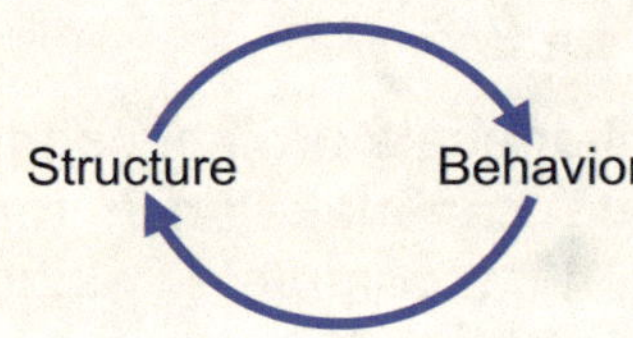

Traffic and roads go together – how do they influence each other? What structures could be put in place that would relieve traffic crowding? By expanding roads, do we set up the situation "if you build it, they will come?" New roads encourage and attract drivers, and soon capacity is exceeded again.

Widening the road didn't seem to make commuting by car easier. How could planners reduce the number of cars traveling during rush hours? Complex problems seldom have simple solutions, and systems usually push back against change. Modeling and simulating can raise and test "what if" questions.

Would more people use buses if a highway lane was designated exclusively for them? A model might show how many cars could be removed from the road by that policy.

Bringing the Lesson Home

How did the systems thinking skills help you think about the problem you were given?

Each team should share how their use of the skills guided their thinking. Students will recognize that complex problems like traffic congestion usually have no simple solutions. Most metropolitan areas must contend with traffic problems. The thinking skills expand our viewpoint and help us consider complex problems from a variety of perspectives. Experience shows that building more road capacity does not eliminate traffic jams – in fact, the situation frequently gets worse after any initial improvement. New roads can trigger vicious cycles by attracting more drivers, causing more congestion, and calls for more road construction.

Systems thinking helps explain counterintuitive outcomes and provides insight into hard-to-understand behavior, such as bigger roads causing increased congestion. Instead of assuming that increasing the capacity of roads will solve the problem of rush hour congestion, systems thinking skills encourage expanding the range of possible strategies.

What problems do you think would be easier to understand using the systems thinking skills you studied or learned about from other teams?

Encourage students to find examples in the world around them. Their school and classroom are complex systems with problems to solve. Looking at the day's news would certainly present a multitude of complicated issues. Students may raise environmental issues, which is the focus of the lessons in this book.

How can we move away from policy structures like road expansion that seem to create more problems?

It is not easy to make changes that are fundamentally different from the way things have been for many years. Political and social institutions are often at odds with environmental goals, but change is possible. Students may be aware of success stories involving change over time. Here are four examples:

- *eliminating smoking in public buildings and reducing tobacco use overall*
- *establishing a system of national parks to protect valuable land*
- *increasing the populations of eagles and other raptors by banning DDT use*
- *expanding legal protections for broader segments of the population*

In a democracy, people can determine policies. It's important for students to be aware of their responsibilities to be "systems citizens." Using the systems thinking skills helps everyone make informed choices that can move the world toward a healthy, sustainable future.

1.3 Electrical Energy – Reducing Carbon and Moving to Renewable Sources

Environmental Systems
How can we produce efficient, non-polluting energy?

Tools

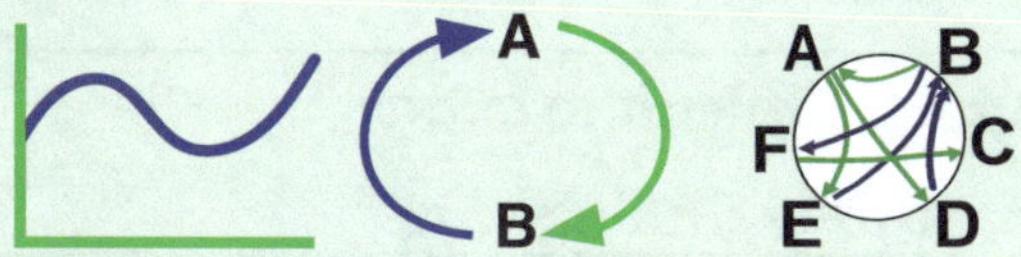

Skills

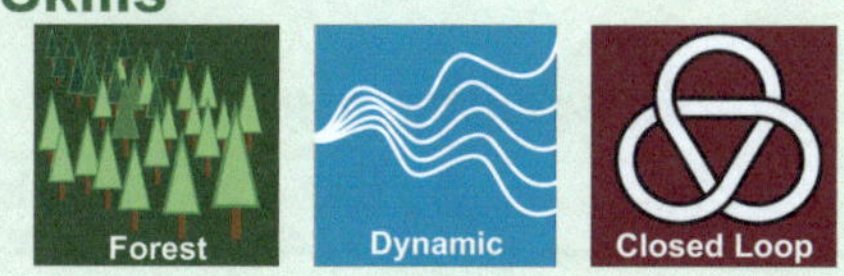

"The least polluting use of energy is not using energy." Bernard Laponche

Overview

Electricity use has transformed the world throughout the 20th and 21st Centuries, improving the standard of living for billions of people. But more than 50% of electricity is generated by burning fossil fuels, causing pollution and driving climate change. Reducing fossil fuel use will be difficult, and developing any source of energy involves trade-offs.

Students examine the costs and benefits of changing from a carbon-based energy grid and heading toward renewable sources of energy. They study graphs that chart electricity production and usage and read about promising initiatives. Tracing feedback through causal loop diagrams, they analyze the short-term and long-term outcomes.

Materials

- Graphs on electricity use over time, p 198
- Suggested articles, p 198
- Handout: Energy Types, p 31
- Handout: Energy Loops, p 32
- Resource: Intro to BOTGs, p 215
- Resource: Intro to Connection Circles, p 218
- Resource: Intro to CLDs, p 220
- Resource: Intro to SD Modeling Software for drawing diagrams (optional), p 238

Vocabulary and Tools

Behavior over time graphs are line graphs that track one or more variables over time. By convention, time is measured along the horizontal or x-axis and variables along the vertical or y-axis. Multiple variables may be tracked on one graph pad.

Connection circles are diagrams that trace causal connections among the elements or parts of a system. Connection circles may be used as a brainstorming tool to build causal loop diagrams. Follow causal arrows to find closed loops.

Causal loop diagrams are single or interconnected feedback loops that illustrate how elements in a system affect each other. There are two kinds of feedback loops – Elements in **reinforcing** feedback loops keep changing in the same direction, while elements in **balancing** feedback loops oscillate, or reverse their direction of change over time.

See the Appendix C for more detailed explanations of these system dynamics tools and links to further learning.

Procedure

1. Ask students to study graphs of electricity use over time. Use the suggested resource or have them research their own sources.
2. Before they begin researching, have teams use the included handout to create a table listing ways electrical energy is generated. Encourage them to brainstorm the advantages and disadvantages of each method. Here are some examples:

TYPE	ADVANTAGES	DISADVANTAGES
Coal	Abundant Relatively inexpensive	Polluting, both its extraction and combustion
Oil	Portable Produces much energy per unit	Polluting, difficult to extract Finite supply diminishing
Solar	Clean Renewable No recurring fuel charge	Intermittent Seasonally variable, weather dependent Requires large area
Hydropower	Clean Renewable No recurring fuel charge	Limited sites Displaces people, alters landscape and waterways Precipitation varies
Wind	Clean Renewable No recurring fuel charge	Inconsistent Large structures impact environment and wildlife
Biofuels	Renewable	Reduces agricultural land for food production Energy intensive to produce Emits carbon
Geothermal	Clean Renewable No recurring fuel charge	Unevenly distributed on Earth Requires high temperatures
Hydrogen	Clean, renewable Potentially abundant	Energy intensive to produce
Nuclear	Little CO_2 emitted Large amount of energy produced per volume of fuel	Expensive to build Dangerous waste products Potential for accidents Fuel expensive to extract, process, store, and dispose

3. Vaclav Smil believes that replacing the dominant energy sources of coal, oil, and natural gas becomes harder as the overall demand for electricity increases. Have students read "The Long Slow Rise of Solar and Wind" or other materials to learn about some of the obstacles to shifting away from fossil fuel use.

   Increasing the supply of non-fossil, renewable energy certainly presents a difficult challenge, but there are examples of progress, even if there are no "magic bullets" or "free lunch."

   Have teams of students investigate a variety of initiatives and create feedback loops explaining each case. Use the articles linked in the materials section or have students search for other reports.

   When teams of students share their diagrams, common characteristics or generic structures should become evident.

   Examples of alternative energy generation from the linked articles are listed below with sample causal loop diagrams. Generic causal loop diagrams about energy follow in #4.

   Trust students to create their own feedback loops. Use the examples as guides for coaching. Be sure students are able to explain the causal connecting arrows, either by citing parts of the article, research they have done, or assumptions they are making based on prior knowledge.

   Article: "The Power Plants That May Save a Park, and Aid a Country" (FIGURE 1)
   Successful hydroelectric projects spur economic growth in the Democratic Republic of Congo. Support for national parks and tourism grows while violence and conflict are reduced. Fewer trees need to be cut down for use as charcoal.

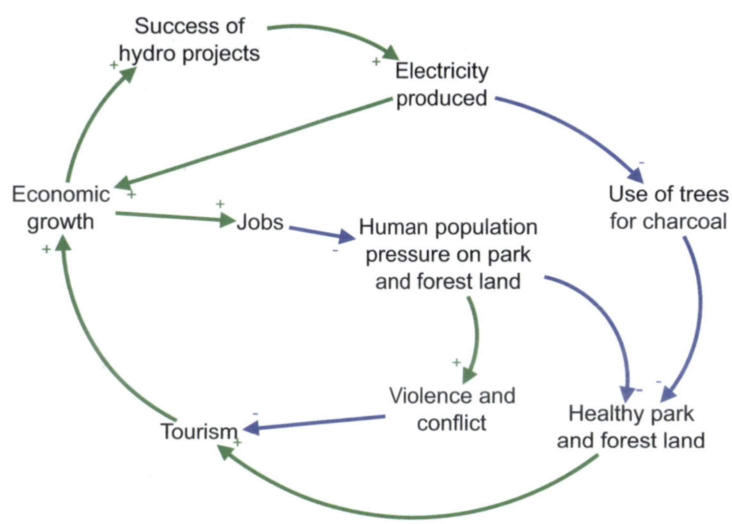

FIGURE 1

   Article: "Elon Musk Just Built a Battery the Size of a Football Field in less than 100 Days" (FIGURE 2)
   A giant battery was built to store surplus electrical energy. Intermittent sources of energy like wind and solar can become more reliable and consistent.

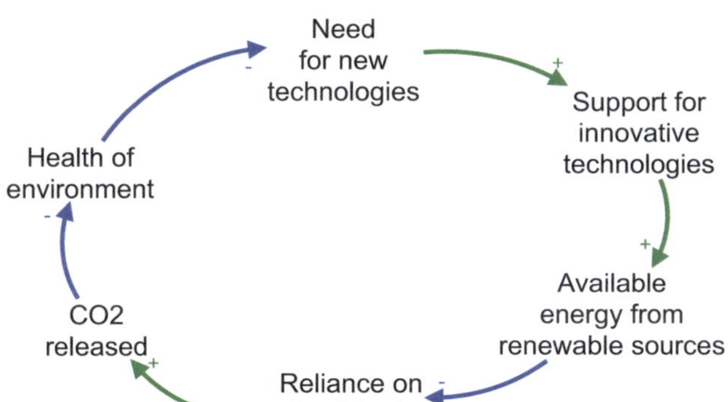

FIGURE 2

Article: "Chile's Energy Transformation is Powered by Wind, Sun and Volcanoes" (FIGURE 3)
Investing in a grid of clean energy sources helps to avoid the shortcomings of each type.

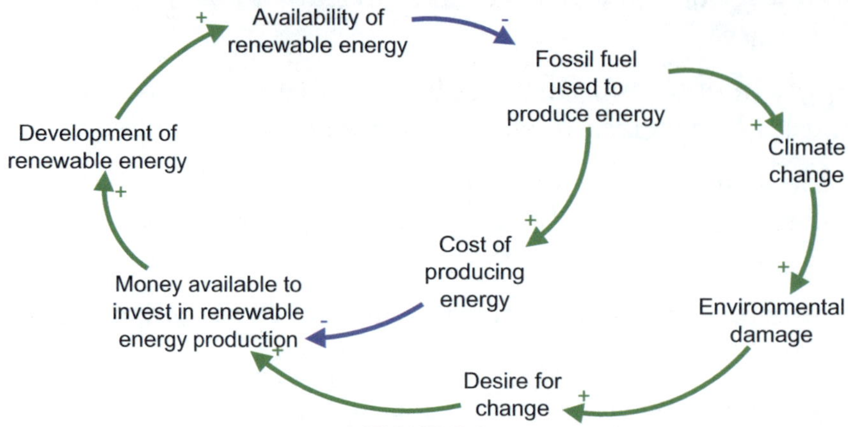

FIGURE 3

Article: "The $3 Billion Plan to Turn Hoover Dam Into a Giant Battery" (FIGURE 4)
Water that has traveled through turbines and produced electricity may be pumped back to a reservoir upstream using surplus solar energy.

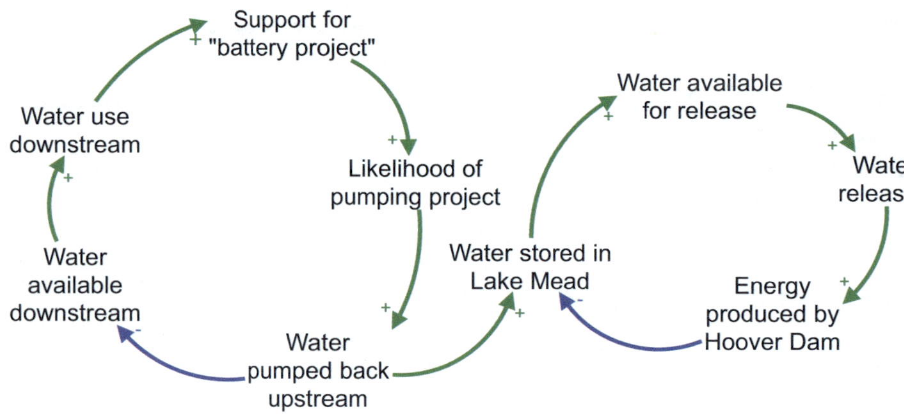

FIGURE 4

Article: "The Biggest, Strangest 'Batteries'" (FIGURE 5)
Because they can be stored and then burned when needed, fossil fuels can produce enough energy for periods of high demand. Engineers are inventing a variety of systems to store energy in other ways.

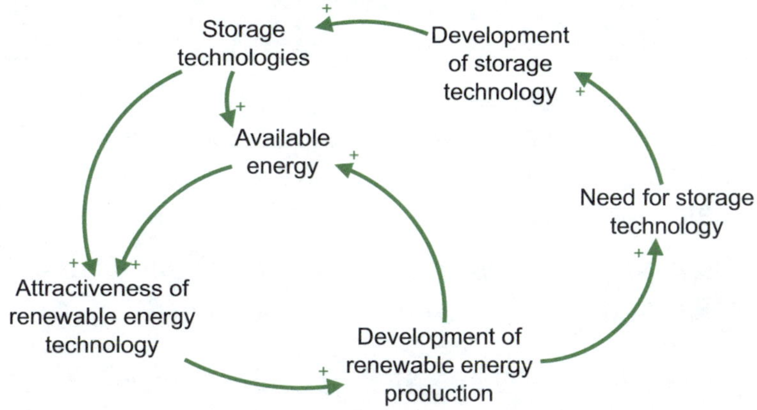

FIGURE 5

4. Examples of causal loop diagrams illustrating generic structures, i.e., a set of interdependencies that often occur across multiple settings and/or contexts:

   Economic activity and energy growth drive each other in a reinforcing loop (FIGURE 6).

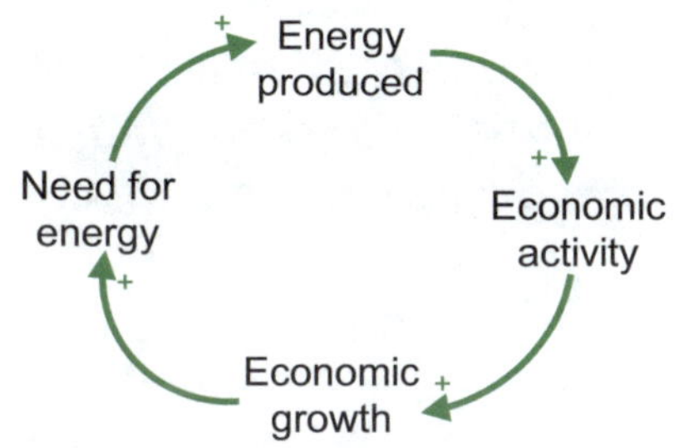

FIGURE 6

   Energy production brings pollution. When consequences become severe, people need to develop alternative sources of energy. Sometimes significant delays occur. Delays are represented in causal loops with double lines across the connecting arrow, indicating a causal connection that takes more time relative to the other links in the loop (FIGURE 7).

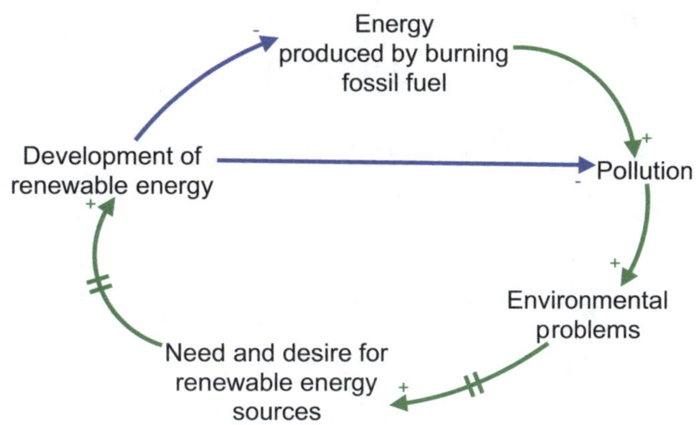

FIGURE 7

Bringing the Lesson Home

What was successful in the case you studied? What problems were encountered?

Each team will have different stories to tell. Look for common themes: politics and economics always influence policy decisions. All methods of generating energy require some natural resources and space that could be used in other ways:

- *Hydroelectricity – Dam building affects ecosystems and recreational use. Areas behind dams are intentionally flooded, displacing people and wildlife.*
- *Batteries – Chemical materials must be mined, are expensive, and sometimes toxic*
- *Storage – "Batteries" like Hoover Dam are large and expensive. Most infrastructure projects require enormous investment and can require very large areas of land.*
- *Solar – Materials must be mined; collectors take up space.*
- *Wind – Farms can be large scale, noisy and intrusive to people, and dangerous for wildlife.*
- *Geothermal – Projects can damage the environment.*
- *Biofuel – Usage releases CO_2 when burned and takes up cropland that could produce food.*

Why are there long delays in changing the way we produce electricity?

The energy grid is enormous and extremely complex. Companies that produce electricity have huge investments in current infrastructure and may be resistant to change. Building new infrastructures requires time and large investments of capital. Fossil fuels are convenient and entrenched in our societies. They contain high amounts of energy per unit of mass, are transportable and storable, and can be burned on demand. Many renewable, cleaner sources of energy are intermittent and difficult to match supply and demand.

How does improving energy efficiency address some of the world's problems?

Everyone in the world uses energy. Increasing access to clean, affordable electricity will have far reaching effects because so many problems are connected. Electricity can drive economic development, which raises people's standard of living and their overall well-being, but often has negative effects on the environment. Energy production from fossil fuels changes Earth's climate and diminishes resources over time. Clean, efficient energy can address these problems and others.

What do you see as helpful steps toward a future world that uses renewable sources of energy?

Students will cite advances they discovered in the research they did. Multiple strategies need to be pursued since it is unlikely that a single technological advance by itself will move the world to a cleaner energy future.

- *Developing a range of appropriate technologies for each region rather than relying on one energy source*
- *Leveling off of demand through conservation, efficiency, and slowing rates of population growth*
- *Improving technology in clean energy production, storage, and transmission*
- *Balancing several goals at the same time – high standard of living, clean and healthy environment, and peaceful and cooperative relationships within and among nations of the world*

Handout - Energy Types

TYPE	ADVANTAGES	DISADVANTAGES
Coal		
Oil		
Solar		
Hydropower		
Wind		
Biofuels		
Geothermal		
Hydrogen		
Nuclear		

Handout - Energy Loops

Article:	
Main Initiative:	
Connection Circle:	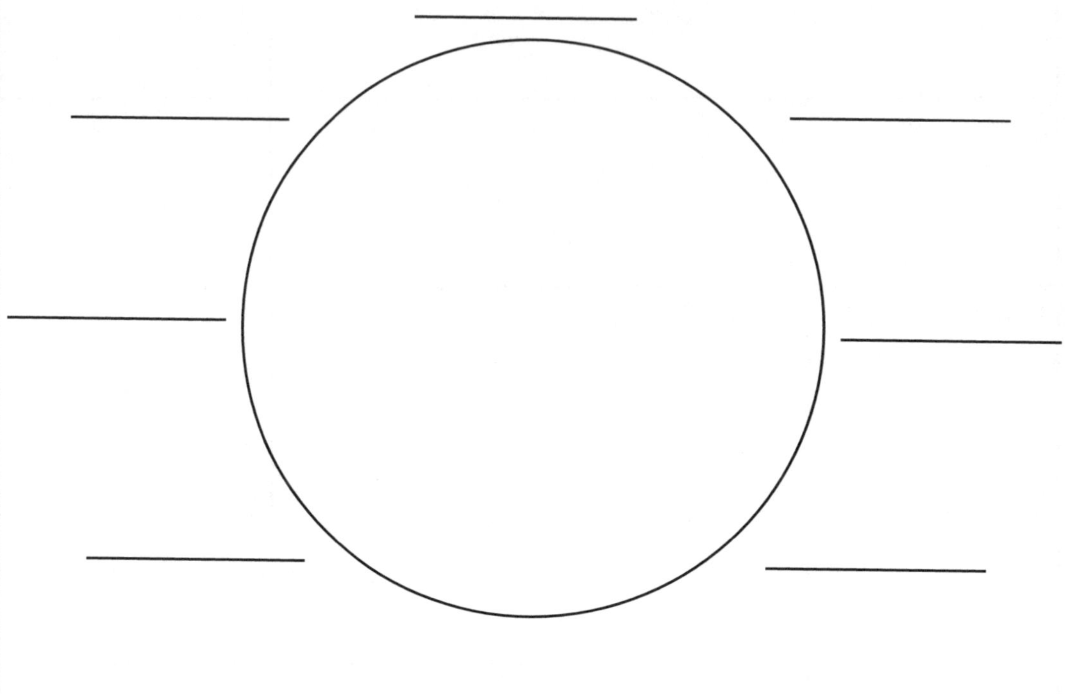
Feedback Loop(s):	
Conclusions:	

1.4 Recycling – When We Throw Stuff Away, Where is "Away?

Environmental Systems
- How does recycling change the flow of waste disposal?

Tools

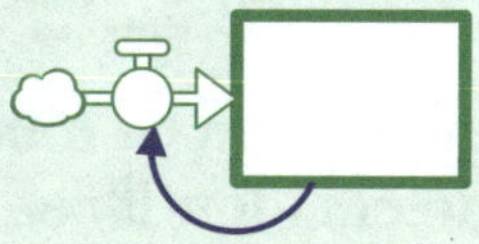

Skills

"There is no such thing as 'away.' When we throw anything away it must go somewhere." Annie Leonard

Overview
What happens to products when their useful life is over, or consumers want to replace them with newer goods? Students investigate how recycling can help build a sustainable society. Using stock/flow diagrams, they model how reducing or eliminating waste has many positive outcomes.

Stock/Flow Diagrams and Models
When stocks and flows are used to describe situations, the resulting graphic representations can be called diagrams, maps, or models. A drawing of the structure of a system creates a "diagram." Those diagrams describe change over time, and are sometimes referred to as "maps," especially when they have multiple stocks.

When equations and quantities are added, and a simulation can be run (by hand, or more commonly, on a computer), we have created a "model."

Materials
- Suggested articles, videos, and research materials, p 199
- Handout: Advantages and Disadvantages of Recycling, p 38
- Resource: Intro to Stock/Flow Diagrams, p 223
- Resource: Intro to SD Modeling Software to draw diagrams (optional), p 238

Procedure
Part 1: Introduce stock/flow diagramming

1. Introduce stock/flow diagrams using the following analogy of a bathtub. If you use computers to draw the diagrams, have students access the modeling program by creating or logging into their accounts. Otherwise, have them use pencils and paper to sketch their diagrams.
2. Water in a bathtub is a stock, or accumulation. Like all stocks, it is a quantity that can increase, decrease, or stay the same over time. The water entering the tub from the faucet is an inflow, while water leaving the tub is an outflow. Both flows can be regulated, affecting the stock of water in the tub. Students' diagrams should look similar to FIGURE 1.

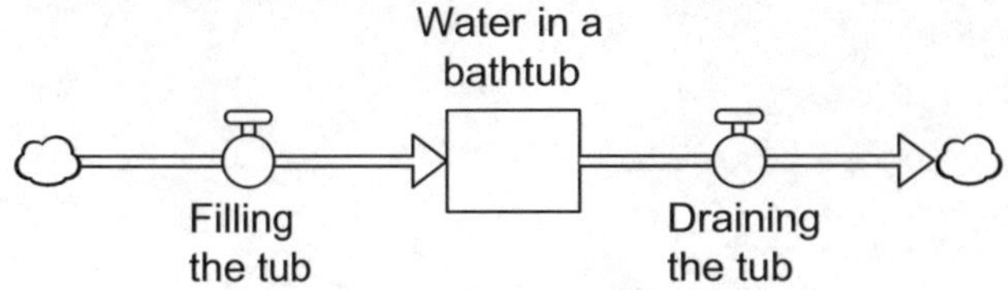

FIGURE 1

3. Explain that the "clouds" on the ends of the flows mark the boundaries of the part of the system being considered. While it is important where the water comes from and where it goes, our diagram is only representing the behavior we have drawn.

4. Ask for examples of other stock and flow situations. Because the same structure can be found in countless situations, have students give their own examples. Many modelers identify stocks first and then connect flows.
Important - the units of the flows must match the units in the stock. Whether the stock is water (FIGURE 1), customers (FIGURE 2), or coins (FIGURE 3), there must be unit consistency.

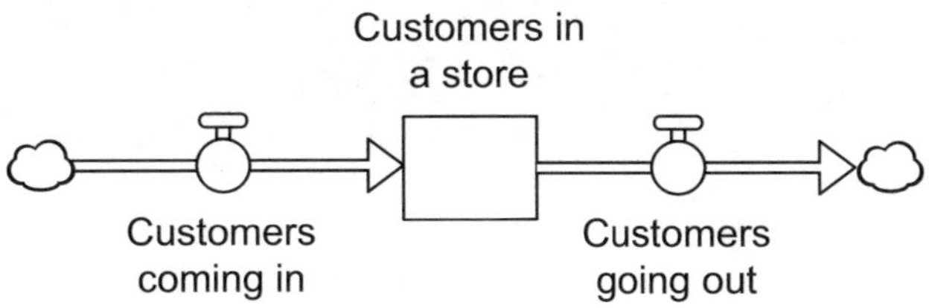

FIGURE 2

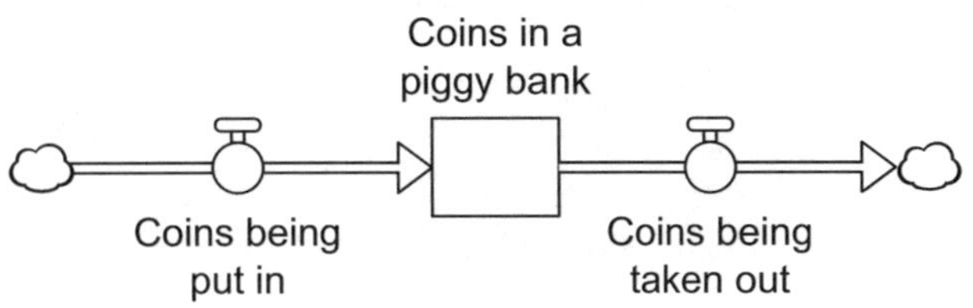

FIGURE 3

Storytelling

Stock/flow diagrams can be especially effective when their structure is revealed one element at a time rather than displaying the completed diagram all at once. Begin with a stock or two, then show the flows that increase and decrease them. Allowing the viewers to see the diagram unfold guides them to understand the logic behind the model builder's thinking. Some software includes built-in storytelling features for unspooling a diagram. Otherwise, reveal the parts sequentially as a series of slides or panels, adding new elements each time.

5. More complicated stock/flow diagrams can have multiple stocks, and stocks may have more than one inflow or outflow. Here's an example of a diagram a high school administrator might make to track the plans of the members of a yearly cohort (FIGURE 4). Divide the graduates into three categories:
   - those who find work (including military service),
   - those who enroll in college, and
   - those who either delay their decision or do something else, such as volunteer, travel, or enter a training program.

Have students draw a stock/flow diagram of their choice to practice.

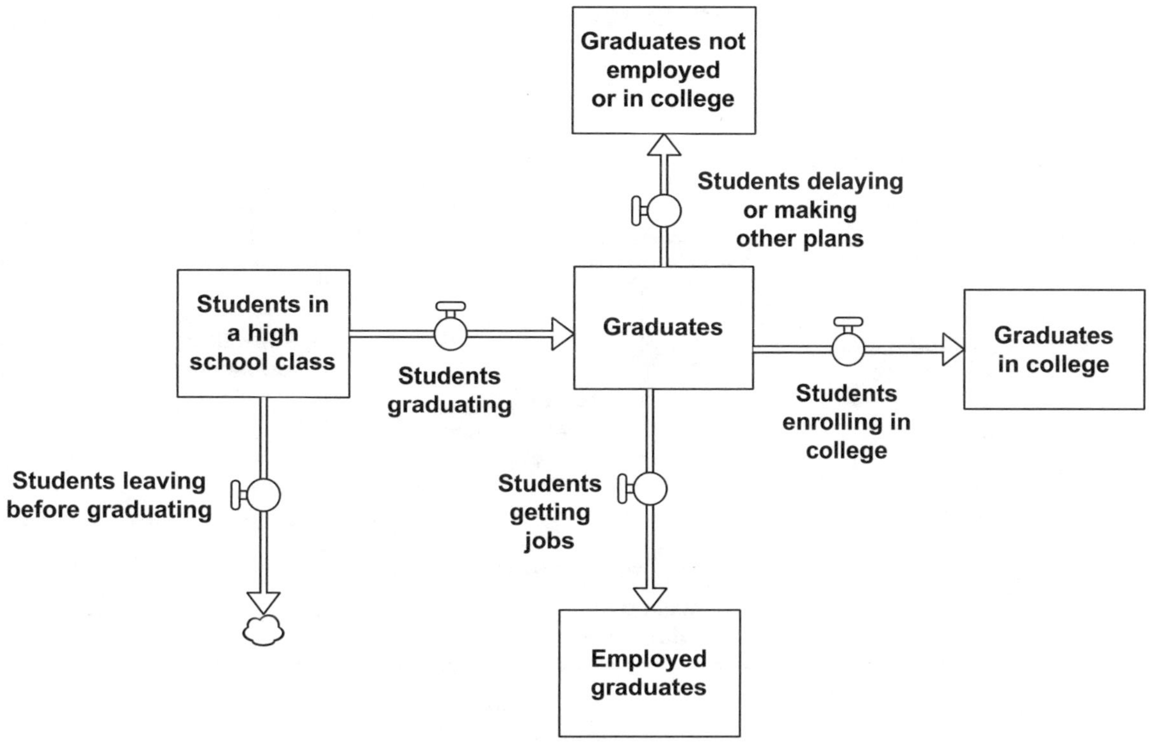

FIGURE 4

Part 2: Research recycling

6. Organize students into pairs or teams, depending on how your class activities are structured. Assign one or more articles about recycling to each team. Use links referenced in Materials, or encourage students to research on their own. Encourage them to look for stocks and flows in the recycling process.

7. Ask students to create a stock/flow diagram of the recycling process. Diagrams will differ, especially if students read different articles.

   Here is one way to represent recycling with stocks and flows (FIGURE 5). Follow the flows to see how materials are recycled and how they can end up as waste. Note how the stock, "Waste materials" lacks an outflow in the diagram. What does students' research tell them happens to waste? Where does it really go in the end?

Some of the steps are simplified, but the diagram shows that both consumers and manufacturers determine how materials travel through the process of manufacturing, use, and disposal.

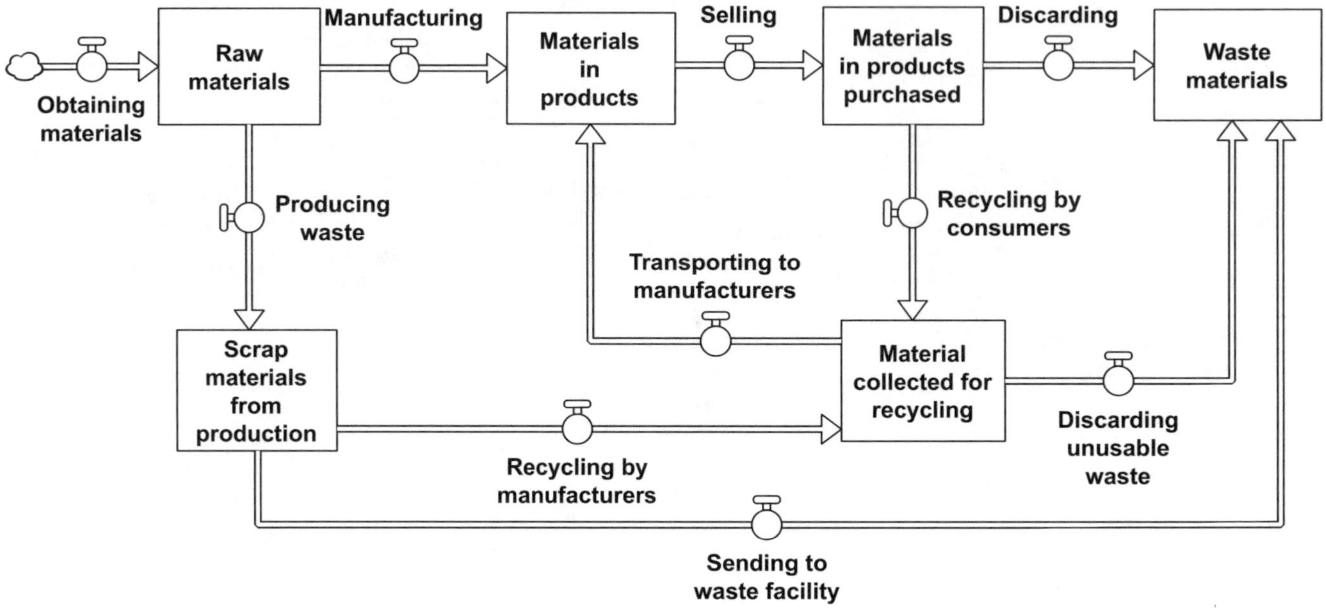

FIGURE 5

Bringing the Lesson Home

How did your team construct your stock/flow diagram about recycling and waste disposal? Did anything you learned surprise you?

Some teams will have identified stocks as the stages where materials accumulate. Others may have chosen a product and traced its path. Be sure students share what important information they learned through their research.

What are the advantages of recycling? What are the disadvantages? (See included handout.)

Among the advantages of recycling:

- *Reduces the use of natural resources*
- *Keeps materials out of landfills, incinerators, and the environment*
- *Can be less expensive and cause less pollution than producing products from raw materials*
- *Saves energy*
- *Creates economic opportunities*

Among the disadvantages:

- *Recycled materials may be difficult to sort and contaminated by non-recyclable items*
- *Quality of materials may be poorer than new ones*
- *Leftover waste from recycling can be difficult to handle*
- *Facilities for recycling require space*
- *Recycling can be more expensive than using raw materials*

Handout: Advantages and Disadvantages of Recycling

Advantages	Disadvantages

1.5 Environmental History
Important Events from the Past

Environmental Systems

• What led to critical environmental events of the past, and what can they teach us?

Tools

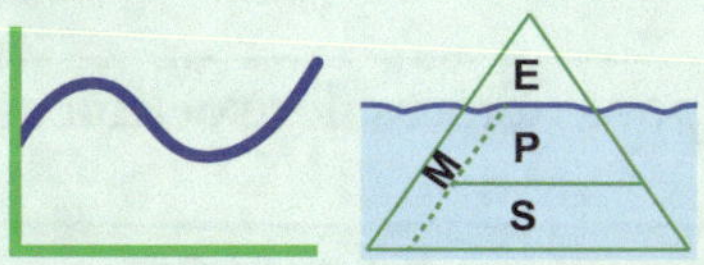

Skills

"Hindsight is 20/20."
Richard Armour

Overview

Students research events and crises that have changed environmental history. They include two systems thinking tools while reporting: behavior over time graphs and iceberg diagrams. In the process, they gain historical context for how the past has led to current conditions, and also how mental models underlie structures that cause system behaviors.

Materials

- Resource materials about important cases in environmental history
  See suggested list of topics (FIGURE 1) and find information online or in print. Specific articles and videos will depend on individual student and teacher choice. Factors to take into account:
  - ◊ Degree of teacher guidance for student choice concerning type of environmental incident, location, period of history, etc.
  - ◊ Student reading proficiency and prior knowledge
  - ◊ Individual interest in topic
- Handout: Topic Ideas, p 48
- Handout: Behavior Over Time, p 217
- Handout: Template for iceberg diagram, p 227
- Resource: Intro to BOTGs, p 215
- Resource: Intro to the iceberg, p 226

Procedure

1. Before class, prepare the template for iceberg diagrams to use with students. Then compile a list of critical events in environmental history. Choose topics you feel are important for students to learn and that they can research online or in print resources.

An online search will find timelines of events. Here are some important ones beginning in the early 20th Century (FIGURE 1). Sample iceberg ideas are included after the debriefing section, Bringing the Lesson Home.

Year	Event	Description
1900	Galveston hurricane	Powerful hurricane devastates a city with rain, wind, and tidal surges.
1906	San Francisco earthquake	Strong earthquake destroys buildings and causes fires.
1930s	Dust Bowl	Intense dust storms during periods of drought decimate agricultural lands.
1940s on	Nuclear weapons	Atomic fission produces weapons of mass destruction and lasting toxicity.
1962	*Silent Spring* published	Scientist writes about the consequences of pesticides.
1978	Love Canal	Chemical dumpsite pollutes neighborhood.
1978	Three Mile Island	Meltdown and radiation leak in nuclear reactor
1984	Bhopal disaster	Gas leak at a chemical plant
1986	Chernobyl meltdown	Explosions and fires destroy nuclear power plant.
1989	Exxon Valdez oil spill	Oil tanker runs aground and spills massive quantity of oil into the sea.
1991	Kuwaiti oil fires	Military forces set fires during war.
1993 on	US Midwest floods	Heavy rains cause river flooding.
2005	Hurricane Katrina	Powerful storm causes many casualties and much damage.
2010	Deepwater Horizon oil spill	Off-shore oil drilling rig explosion releases huge quantity of oil into sea.
2011	Tsunami in Japan	Earthquake and tsunami cause massive loss of life and destruction, including a nuclear power plant.
2012	Hurricane Sandy	Destructive deadly storm
2017	Series of hurricanes	Succession of disastrous storms
2019 on	Wildfires around the world	Devastating fires fueled by drought, high temperatures, and wind

FIGURE 1

2. Tell students that they will work in teams to research and report on important events in environmental history. Have students compile a list or use the included handout, so they can choose topics that interest them. Assign teams before events are chosen, or have students submit a list of their topic choices and group them by interest.

   Distribute copies of an iceberg template, or project an image in the classroom.

3. When teams and topics have been chosen, tell students that their presentations must include examples of two systems tools:

   - Behavior over time graphs (BOTGs) – Choose 2-4 variables to track over a period of time that includes the environmental crisis they are studying
   - An iceberg diagram – Create a poster describing their topic through three levels of an iceberg diagram, with mental models noted also. Iceberg layers may contain systems diagrams, such as BOTGs, causal loop diagrams, and stock/flow diagrams.

Procedures #4 and #5 provide a quick overview of each tool and how they can add insight to students' work.

4. BOTGs are line graphs with the time scale on the horizontal or x-axis. The vertical or y-axis scale measures some attribute of a variable being tracked. As this example shows, multiple variables can be placed on the same graph pad. Using the same time scale, students can even graph variables with different units noted on the y-axis.

   The graph (FIGURE 2) shows that nuclear energy is a significant source of energy but by 2015 had been surpassed by renewable energy sources.

   Even though we may call these incidents in environmental history events, they did not occur in isolated time periods. What came before, what happened afterward, what was happening over a period of time? All these questions are critical to ask when investigating a complex system. Systems thinking principles and tools, like BOTGs, encourage looking at history, current conditions, and future projections. BOTGs display continuous patterns of change rather than points of discrete events. While students will discover specific data points for their graphs through research, encourage them to look at the overall shape of the graph to analyze change over time.

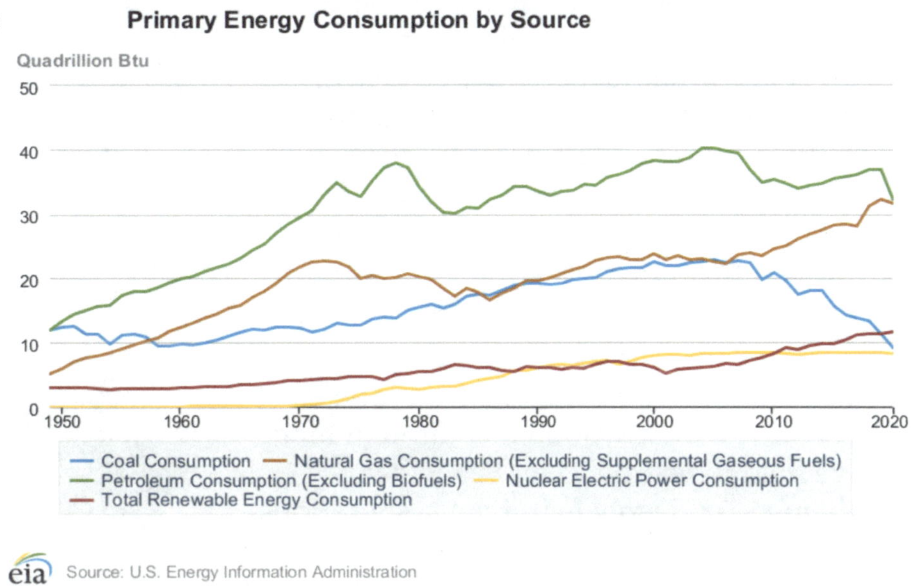

FIGURE 2

   For each behavior over time graph, have students write or explain brief answers to four questions:

What is changing?	Identify the variables on the graphs. Explain how the data was measured and other details that are important.
How is it changing?	Describe the rate of change shown by the shape of the graph.
Why is it changing?	Tell why you think (based on your research) the variables behaved as they did.
So what and now what?	What is the significance of the graph? Decide what policies or decisions these data might require.

5. Have each team create an iceberg diagram (FIGURE 3) to accompany their presentation. If students are unfamiliar with the tool, explain that an iceberg diagram has three layers. Like an iceberg in the ocean, what is visible is a small percentage of the total. Here's a brief description of each iceberg layer, followed by an example using a nuclear power accident:

Event Level

The top layer is what we see, i.e., occurrences that we observe happening. In this activity, that will be the environmental events.

Patterns Level

Just under the 'water,' this level looks at a broader time frame. What series of events or trends over time have led to the event? Students' BOTGs will fit this layer, along with descriptions of conditions leading up to an event.

Structures Level

In this layer, students analyze causality. What is causing the patterns in the layer above? Students should discover cause and effect relationships from their research and possibly draw feedback loops to illustrate those structures. If they have created stock/flow diagrams, include them here also.

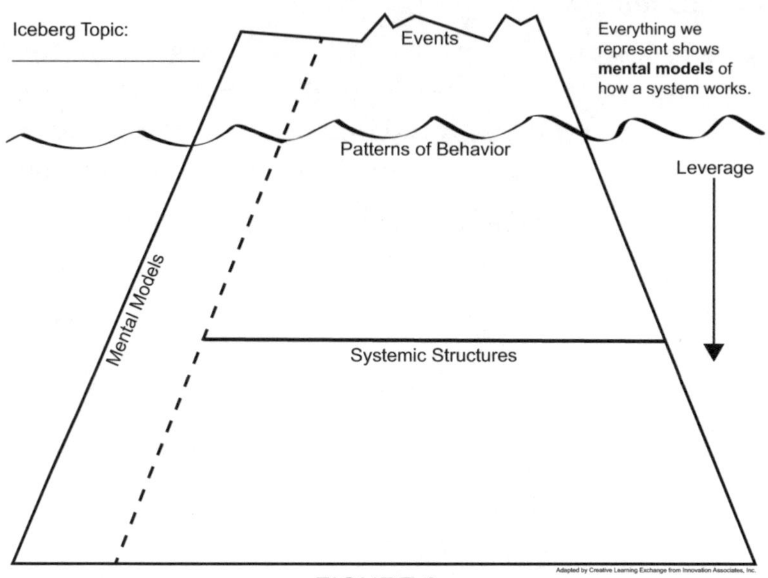

FIGURE 3

Mental Models

These beliefs and assumptions are integral to the layers of the iceberg. Everyone constructs their world view through models we hold to be true. Peter Senge defines them as …"deeply ingrained assumptions, generalizations, or even pictures or images that influence how we understand the world and how we take action. Very often, we are not consciously aware of our mental models or the effects they have on our behavior."

Encourage students to use the iceberg as an organizing graphic for their presentation. Suggest that they ground each layer in research as much as possible.

The iceberg diagram (FIGURE 4) demonstrates how the study of a nuclear accident at a power plant might be organized. While this example has text only, students will include diagrams as much as possible.

Types of Iceberg Diagrams

Many systems thinkers consider mental models to be inherent in all levels of our thinking. For example, if we create a graph or causal loop, we are including our mental models in the diagram itself. Mental models also influence each of the other three levels of the iceberg. In this activity and the rest of the book, we draw additional mental models along the side of three-level iceberg diagrams.

Some systems thinking practitioners prefer to use icebergs with four levels, reasoning that mental models are so deeply ingrained in our thinking that they belong at the deepest level. Others use a five-level diagram, explaining that mental models are formed and supported by an even deeper level of values and beliefs.

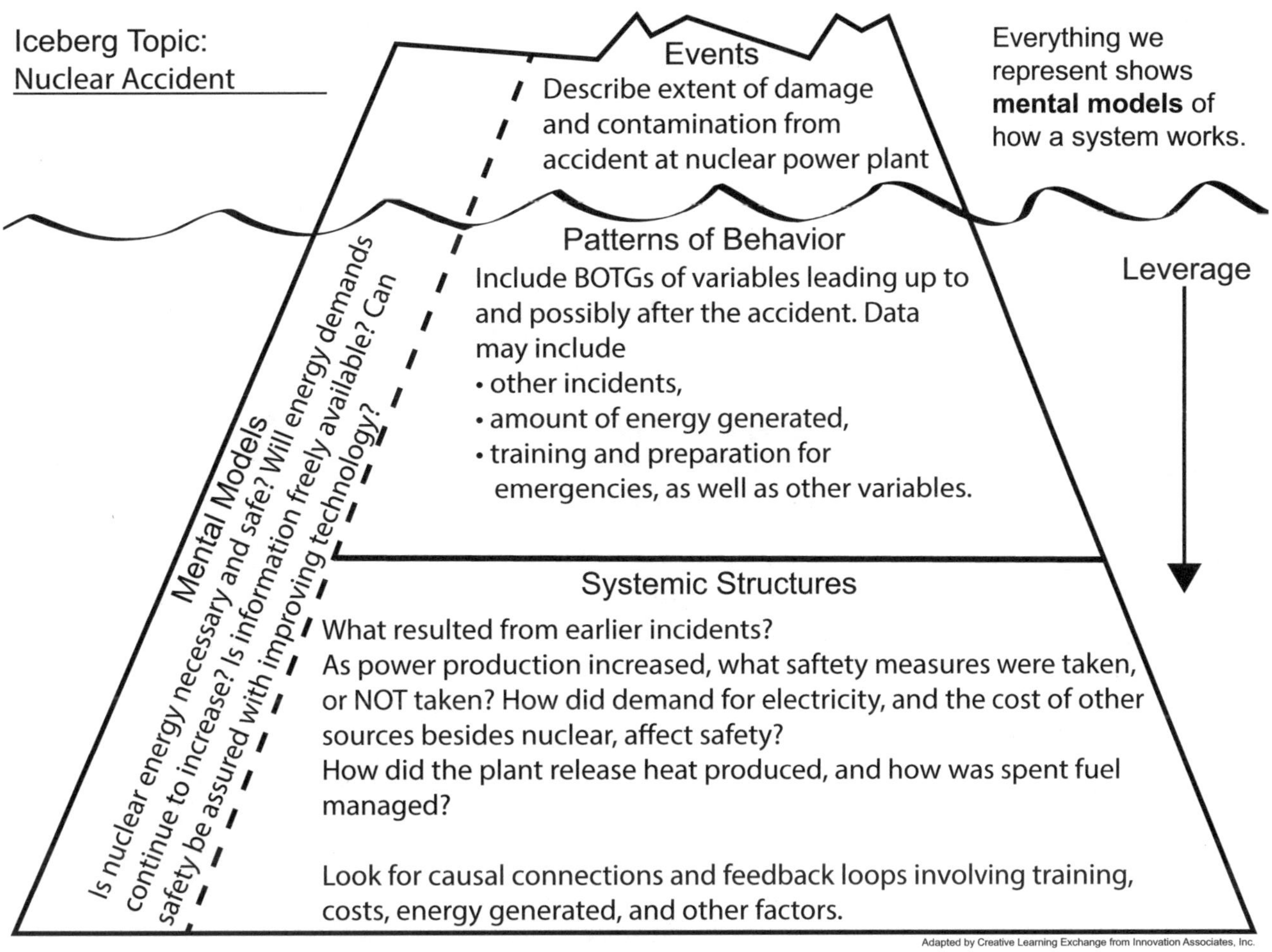

FIGURE 4

Interdisciplinary Learning

When studying environmental systems, students will frequently cross traditional course boundaries. History and science obviously intersect in this activity, and graphing involves math skills and quantitative reasoning. Economics is always a factor in environmental events. Students need to demonstrate competence associated with language and presentation skills as they organize and report their findings – research skills, writing, structuring an argument or thesis, and building a physical representation of their thinking.

The use of visual tools like BOTGs and iceberg diagrams helps make ideas clear to an audience. Students' slide shows should include graphs and other diagrams. Visual material also enriches writing assignments structured like op-ed pieces or news articles. Preparing a display board requires students to clarify their presentation so that it can stand alone – an important skill, because they won't always be next to their poster to explain their thinking.

FIGURE 5 shows a sample causal loop diagram for possible use in the Structure layer.

Access to information affects public opinion about nuclear power. When public awareness about risks increases, the insistence on competent oversight and planning will also increase. The reliance on nuclear power and the potential for accidents would both possibly decrease.

When the public remains unaware of risks or considers them acceptable, reliance on nuclear power can continue or even grow.

In causal loop diagrams, '+' [plus] indicates positive correlation, meaning "changes in the same direction." '+' [plus] also may be thought of as "adds to." Conversely, '-' [minus] indicates negative correlation, meaning "changes in the opposite direction." '-' [minus] also means "subtracts from." When any element of a causal loop changes from + to -, or vice versa, the whole loop changes.

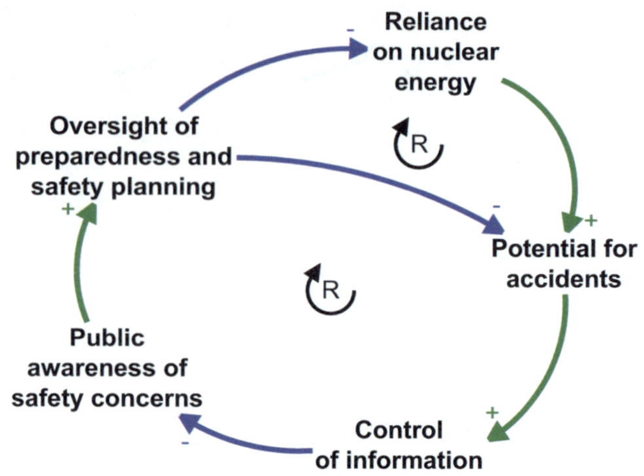

FIGURE 5

6. Examples of iceberg diagram thinking for other environmental events in the table follow the Bringing the Lesson Home debriefing questions.
7. Share presentations within the class or expand to an outside audience. See the sidebar about interdisciplinary learning for suggestions about student options.

Bringing the Lesson Home

How did the BOTGs and iceberg diagrams influence your research?

The BOTGs show how events occur as parts of patterns over time. Looking at a broader picture helps us understand how and why things happen. BOTGs help us ask good questions, which leads to better research in order to find explanations.

Going deeper into a topic with the iceberg diagram leads to fundamental causality and structure. An event is embedded in longer-term patterns of behavior. The iceberg identifies structures that drive behavior, often resulting in unintended consequences. Influencing the formation of the iceberg, deeply established mental models cause people to create and support those structures and patterns that eventually result in important environmental events.

Are there common themes among the environmental history topics?

Students will begin to recognize generic structures common to many of the environmental events they studied. Often business and the environment come into conflict. Humans extract resources from nature and have often been careless with the materials. Some dangerous and destructive products or procedures have been used in ways that are toxic to ecosystems, either before their effects are understood or because of the actions of unscrupulous people.

Bringing the Lesson Home
(continued)

What lessons for the future can be learned from the topic you studied?

Humans must be vigilant to ensure that our actions don't create pollution and other problems. While safety procedures may be difficult and expensive, the costs of environmental degradation are higher and longer lasting. No one wants to sacrifice the convenience and comforts of modern technology, but we have to use care and remain committed to protecting our environment.

What recommendations would you make to avoid or lessen the effects of another event like the one you studied?

Students will have discovered a variety of causes of environmental problems and can recommend policy changes. More careful government oversight combined with citizen education could make corporations less likely to foul the environment. Advances in science and technology should bring cleaner ways to produce energy and materials to support our lifestyles.

How will people become more aware of potential problems like the ones we studied?

Learning about environmental systems will educate people and make citizens demand cleaner and accountable industries. Scientific literacy will lead to more responsible "systems citizens." History has taught many lessons and it is up to us to modify our behavior accordingly.

Examples of iceberg thinking for environmental events

As always, use examples only when necessary, allowing students to do research, form opinions, and create their own mental models and diagrams.

Galveston and other hurricanes

- Events – severe flooding, loss of lives
- Patterns – many previous floods, expansion of city despite threat
- Structures – weak and slow reaction to weather predictions, need for seawall and emergency preparedness plans
- Mental models – belief that city was safe and flooding was rare

San Francisco Earthquake

- Events – much of city destroyed by earthquake and fire
- Patterns – city grew very fast creating dense population and buildings
- Structures – buildings standing on unstable land, failure to prepare adequately for fires, poor understanding of earthquake faults
- Mental models – goal of growth put ahead of safety, poor infrastructure planning

Dust Bowl (FIGURE 6)

- Events – loss of soil, destruction of farms
- Patterns – research BOTGs of rainfall, acres cultivated, number of farms, population
- Structures – reinforcing loop causing soil erosion to worsen
- Mental models – Fertile prairie land can support commercial agriculture; rainfall will be adequate every year.

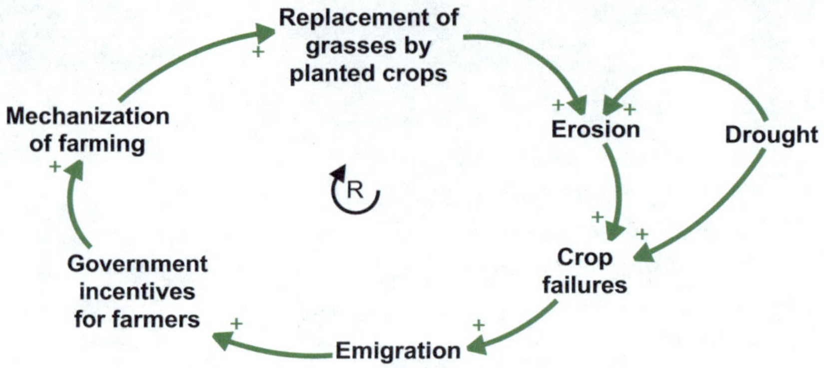

FIGURE 6

Nuclear weapons (FIGURE 7)

- Events – dangerous fallout produced by tests, land destroyed and made uninhabitable
- Patterns – arms race among nuclear powers
- Structures – escalation archetype
- Mental models – Earth is large and fallout will dissipate; the bigger the arsenal, the stronger the deterrent, and the safer a country will be.

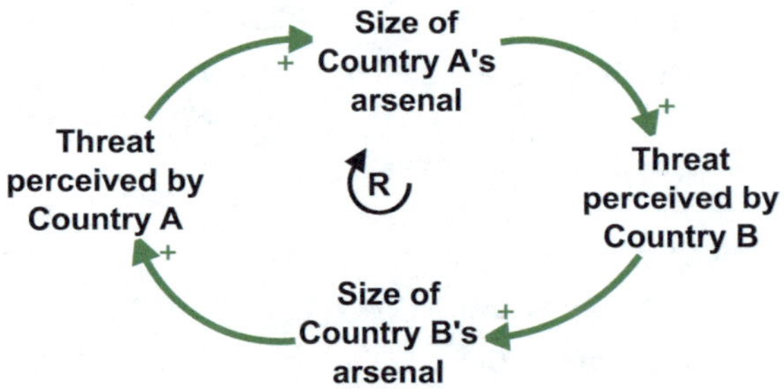

FIGURE 7

***Silent Spring* published**

- Events – pesticide use caused loss of wildlife and increase in illness
- Patterns – BOTGs - use of pesticides, animal populations, rates of illness and death
- Structures – government knows best, women scientists disrespected, chemical industry pushed back on research
- Mental models – Chemicals can increase crop yields by killing pests; as resistance to pesticides develops, stronger poisons are required but will not harm humans.

Love Canal (toxic sites)

- Events – illness in area of chemical plants
- Patterns – BOTGs of rates of illness and death
- Structures – greed of developers putting residents in harmful environments; unsafe practices of waste treatment and disposal; suppression of information
- Mental models – It is safe to build near toxic sites (results in short-term gain at expense of long-term well being).

Bhopal disaster

- Events – toxic release of chemicals
- Patterns – large quantities of pesticides produced; series of leaks at factory
- Structures – poor management practices with dangerous substances
- Mental models – Production of dangerous products usually occurs near areas inhabited by poor people with little political power. Pesticides and other toxic substances are worth the risk and can be managed safely.

Exxon Valdez, Deepwater Horizon (oil spills)

- Events – accidents causing large quantities of oil to be released into the oceans

- Patterns – rising demand and use of oil, production and transport systems loads increasing, leaks and spills over time
- Structures – rising dependence on oil, longer and busier supply lines
- Mental models – safety precautions adequate, okay to ship oil around fragile ecosystems

Kuwaiti oil fires
- Events – war that causes massive oil fires
- Patterns – increasing oil production in politically unstable regions
- Structures – oil resources that cause geopolitical conflicts, deliberate sabotage as acts of war, nations that commit armed forces when disputes arise, uneven distribution of wealth
- Mental models – Corporations rely on adequate protection by military and government against aggression by neighboring countries and organizations using terrorism.

US Midwest floods
- Events – seasonal rain causes rivers to flood, inflicting heavy damage to area
- Patterns – BOTG tracking rainfall and flooding; population density and amount of building and farming on flood plains
- Structures – residential, business, and agricultural enterprises spread onto flood plains; flood control measures overwhelmed; insurance pays for rebuilding
- Mental models – Floods are infrequent. Areas are safe and won't be flooded during residents' lifetimes. Structures like levees will protect low areas.

Tsunami in Japan
- Events – tsunami that floods coast causing loss of lives, property, and destruction of nuclear power plant
- Patterns – BOTG of earthquakes and tsunamis, population living and working in vulnerable area
- Structures – population concentrated near coast, active tectonic areas vulnerable to tsunamis from off shore earthquakes, nuclear fuel dangerous when containment structures are breached
- Mental models – Tsunamis are rare events. Warning system is adequate. Nuclear plant is built to withstand natural threats.

Wildfires (FIGURE 8)
- Events – deadly fires that threaten lives and property, burning across large areas of land
- Patterns – BOTGs of fire frequency and size, rainfall, loss of life and property
- Structures – fire suppression practice that allows build up of fuel, causing larger more dangerous fires
- Mental models – The goal of suppression of all fires makes homes safer and living in forested area safe and desirable.

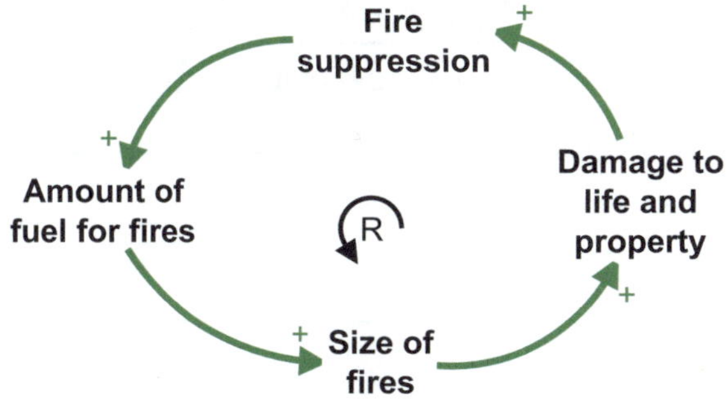

FIGURE 8

Handout: Topic Ideas

Use this chart to identify a topic you're interested in researching. Feel free to think of and add additional ideas at the end.

Year	Event	Description
1900	Galveston hurricane	Powerful hurricane devastates a city with rain, wind, and tidal surges.
1906	San Francisco earthquake	Strong earthquake destroys buildings and causes fires.
1930s	Dust Bowl	Intense dust storms during periods of drought decimate agricultural lands.
1940s on	Nuclear weapons	Atomic fission produces weapons of mass destruction and lasting toxicity.
1962	*Silent Spring* published	Scientist writes about the consequences of pesticides.
1978	Love Canal	Chemical dumpsite pollutes neighborhood.
1978	Three Mile Island	Meltdown and radiation leak in nuclear reactor
1984	Bhopal disaster	Gas leak at a chemical plant
1986	Chernobyl meltdown	Explosions and fires destroy nuclear power plant.
1989	Exxon Valdez oil spill	Oil tanker runs aground and spills massive quantity of oil into the sea.
1991	Kuwaiti oil fires	Military forces set fires during war.
1993 on	US Midwest floods	Heavy rains cause river flooding.
2005	Hurricane Katrina	Powerful storm causes many casualties and much damage.
2010	Deepwater Horizon oil spill	Off-shore oil drilling rig explosion releases huge quantity of oil into sea.
2011	Tsunami in Japan	Earthquake and tsunami cause massive loss of life and destruction, including a nuclear power plant.
2012	Hurricane Sandy	Destructive deadly storm
2017	Series of hurricanes	Succession of disastrous storms
2019 on	Wildfires around the world	Devastating fires fueled by drought, high temperatures, and wind

1.6 Modeling Population How Populations Change

Environmental Systems

- How do populations change?
- How can we model populations?

Tools

Skills

"All models are wrong, but some are useful." George E. P. Box

Overview

Populations change and can create enormous effects on the environment. Every place on Earth feels the impact of humans. Photos from space show our planet as a colorful, gleaming sphere against a black and empty background. Because Earth has finite dimensions, it has a finite capacity to support life.

To get a sense of how populations grow, students play games that demonstrate rates of change. They can build simple computer models of hypothetical populations. By changing elements in the model, they can ask "what if" questions, run simulations, and analyze results based on the outcome.

Materials

- Diagramming and modeling materials
  - ◊ For diagramming, no materials necessary; diagrams may be drawn by hand.
  - ◊ For computer drawing and model building, computers with system dynamics software
- Lesson plans from *The Shape of Change*, p 199
- Population simulations from the Creative Learning Exchange, p 199
- Modeling simulation activity, Bumbles and Woofs, p 199
- Handout: Creating Your Own Population Model (optional), p 57
- Resource: Intro to Stock/Flow Diagrams, p 223
- Resource: Intro to System Dynamics Computer Models, p 224
- Resource: Intro to System Dynamics Modeling Software, p 238

Procedure

Activity 1.4, Recycling, introduces students to stock/flow diagramming. It is not a necessary prerequisite for this activity but may be helpful in building understanding of system dynamics modeling.

1. Optional - To prepare students for modeling, play one or more of the first three simulation

games from *The Shape of Change*. To begin population modeling directly, go to step #3.

- Lesson 1 "The In and Out Game" introduces stocks and flows.
- Lesson 2 "Making Friends" compares linear versus exponential growth, a key concept in population.
- Lesson 3 "The Mammoth Game" demonstrates a standard population model, exactly like the one in this activity, but using a simulated herd of mammoths.

Lesson plans describe the activities and include system dynamics models. Students can build the models, run the simulations, and / or draw the stock / flow diagrams by hand.

2. Optional – For additional experience with population simulations, students can use the activities from the Creative Learning Exchange, listed in Materials.

3. Prepare students to build a stock / flow model of a hypothetical population. Unless you choose to have students draw stock / flow diagrams by hand, set up each individual, pair, or team of students at a computer capable of running system dynamics modeling software. After they open a new modeling screen, encourage students to create and name a stock. Stocks are always nouns or noun phrases such as "Population" or "People in the country." Some modelers follow a convention of capitalizing the first letter of all variables, including stocks.

Stocks and Flows

System dynamics is the quantitative area of systems thinking. Thinking about stocks and flows helps us look at the world operationally, that is, how things work and how things change.

Stocks are accumulations – in a model or diagram, stocks are represented by rectangles that hold quantities of "stuff." In this lesson's activities, the "stuff" is people or animals.

Flows are conduits into and out of stocks, the pipelines that increase or decrease the accumulations. The stuff in the flows has the same units as the stuff in the stocks. If the stock is "mammoths," the inflow will be "births" (mammoths born) and the outflow will be "deaths."

Stock/flow thinking is essential for understanding environmental systems. Stocks are only changed by flows. The relationship between how much flows in and how much flows out controls the quantity in the stock, a seemingly simple concept that isn't always easy to grasp, especially in a complex system.

4. After all teams have placed a stock in their diagrams (FIGURE 1), ask students how populations change. Explain that for simplicity, you will ignore people moving in and out of the population, assuming those numbers will be about equal.

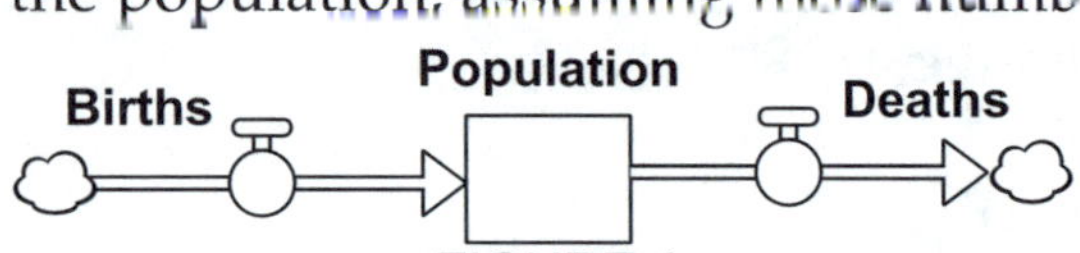

FIGURE 1

Have students attach flows to their stock, being sure the direction of each flow is oriented correctly. Names may vary. A simple label of "Births" and "Deaths" for the flows are sufficient, although phrases like "People being born" are accurate also.

The "clouds" at the ends of the flows indicate boundaries of our diagram. When individuals from the stock of population die, they leave the model through the deaths flow.

5. What controls the number of births in this diagram? Students will be able to explain that larger populations usually produce more births, all other factors being equal. The same holds true for

the deaths flow. In a large city, more people are born and more die per year than in a small town, again assuming no extraordinary events. To indicate the connection between the stock of "Population" and the flows of "Births" and "Deaths," feedback lines called connectors are drawn from the stock to each flow. There is now a visual connection in the diagram indicating two feedback loops (FIGURE 2).

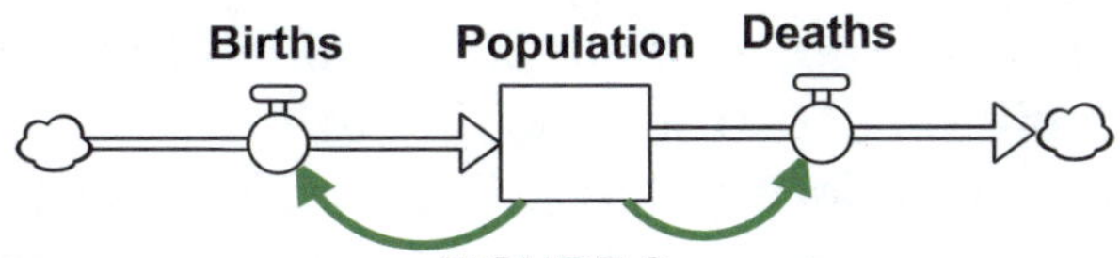

FIGURE 2

Modeling Software

Choosing which modeling software to use will be a matter of personal preference. Some software needs to be downloaded to individual computers while other applications are web-based. Some companies offer free user download or registration, other companies charge a fee. Often, fee-based software will be steeply discounted for educational institutions, so it pays to check with the company if you prefer one that is not free.

Students may identify other factors contributing to the birth and death flows. Ask them to wait to add them to the model, which will happen in step #10.

6. In their teams, or with partners, have students follow the arrows as they trace the two feedback loops. For reference, decide on names for the loops – here we use "Birth loop" and "Death loop."

   First, look at each loop independently. Have teams or pairs of students work together and take time to interpret what the feedback diagrams mean.

7. Birth loop (FIGURE 3)

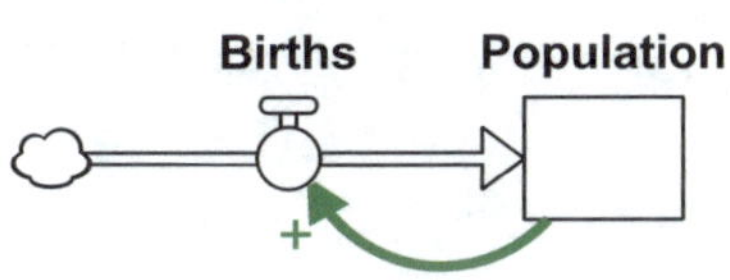

FIGURE 3

The larger the population, the more "Births." The more "Births," the larger the "Population." Because the two variables affect each other in the same way, the birth loop is a reinforcing loop.

8. Death loop (FIGURE 4)

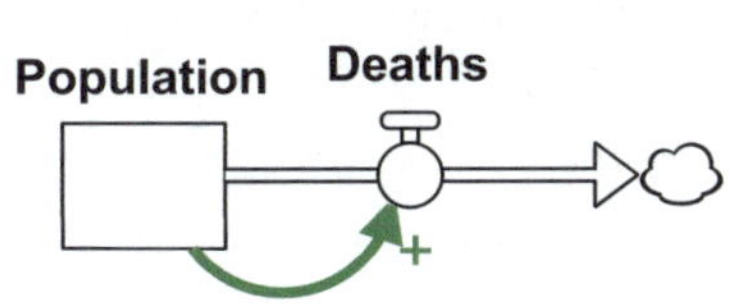

FIGURE 4

The larger the "Population," the more "Deaths." The more "Deaths" that flow out of a "Population," the smaller the "Population." Because of the opposite causal relationship between the elements in this loop, students should recognize it as a balancing loop.

9. Both flows together (FIGURE 5) and then drawn as a causal loop diagram (FIGURE 6):

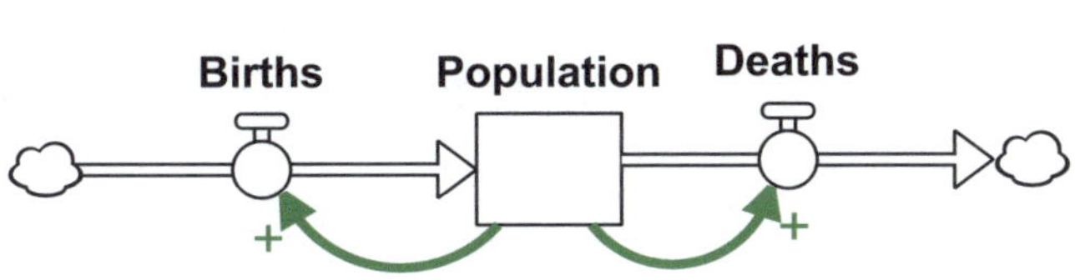

FIGURE 5

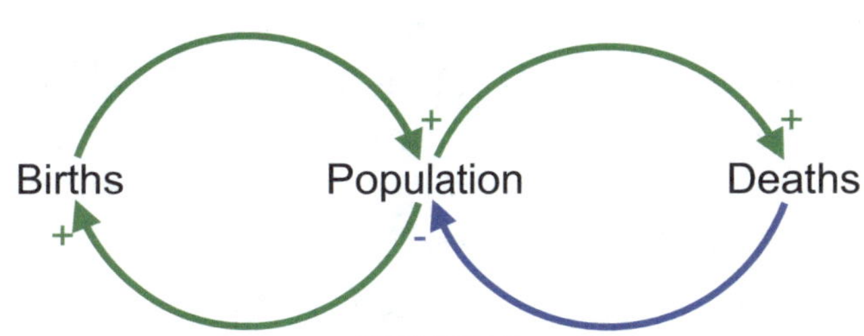

FIGURE 6

When the birth and death loops are combined in one diagram, it becomes clearer to see how stocks and flows can interact. The loops are competing to affect population.

The "Birth loop" is a reinforcing loop, driving the "Population" higher. The more "Births," the greater the "Population," which can then go on to produce more "Births."

The "Death loop" is operating simultaneously, but as a balancing loop. The more "Deaths," the fewer individuals remain in the "Population." That means fewer opportunities for "Births."

According to the diagram, the two counteracting loops would seem to keep a population at some level of equilibrium. How can we account for populations that grow or decline over time? Prompt students to brainstorm why populations change over time. The following examples can provide context:

- Human population has experienced exponential growth, climbing from an estimated three million (3,000,000) hunter-gatherers over 30,000 years ago, to about 300 million (300,000,000) people 2,000 years ago, and to more than seven billion (7,000,000,000) in 2015.
- Conversely, wooly mammoth population in the world declined to extinction. Students will know many other species that became extinct or those currently being severely threatened.

Clearly, the competing birth and death loops do not always produce a steady stock of "Population." Students will generate many explanations for rising and falling populations. The connecting thread will be that when birth and death rates do not match, populations change.

10. Ask students what affects births and deaths and how their rates could be represented and calculated in their stock/flow population diagrams. They will likely reason that the quantity in

Differences between feedback loop diagrams and stock/flow modeling.

In the reinforcing loop containing "Births" and "Population" (FIGURE 6), those two elements move in the same direction. But what happens when population declines for some reason? According to the logic of our reinforcing feedback loop, when Population decreases, births will decline also. Then if births decrease, population should decrease also, according to the causal loop. But that doesn't make sense – in real life, when births flow into population, any births add to population. So what's going on?

Here is where stocks and flows can add insight when feedback diagrams seem illogical. We need to look at this feedback loop as describing rates of increase and use language that accurately tells what happens when the variables begin to decrease. When "Population" goes down, "Births" go down also, and although "Births" add to the "Population," the rate of new individuals entering the "Population" is slower. If more people die than are being born, the "Population" will go down, but births alone cannot decrease a population.

Using precise language to explain reinforcing feedback in an inflow can be a challenge. Students may find it helpful to use the terms "stock" and "flow" in their descriptions. Here's an example:

"When the stock of 'Population' goes down, the number of new 'Births' in the flow per time unit goes down. When there are fewer 'Births' flowing in, that doesn't make the stock go down, but the rate of new individuals entering the 'Population' stock is slower. That means the stock isn't as large as it would have been with a greater flow." (FIGURE 3)

Stock/flow thinking explains that flows control the change in stocks. When there are feedback connections, stocks control the rate of change in flows also.

the flow "Births" results from the number of females giving birth in a year. Many factors determine that rate of flow, including how many females there are, the age of the females, and the likelihood that they will give birth. To keep our stock/flow model simple, we can aggregate, or bundle all factors into one term. This term will represent a name for the fraction or percentage of the population likely to give birth in a year. Solicit suggestions for a name for those factors. "Birth fraction" is a helpful term to use in population models (FIGURE 7). While "Birth rate" is also acceptable and is used more often in everyday language, labeling the term "Birth fraction" for modeling reinforces that a certain fraction of the population gives birth each year, helping students understand the mathematics.

A similar term is needed for the factors affecting the "Deaths" flow. On average, a certain fraction of the "Population" will die each year. We can aggregate those factors into one term called "Death fraction."

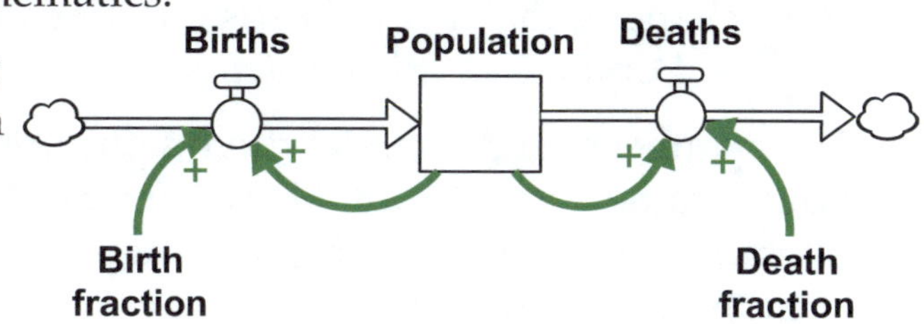

FIGURE 7

In stock/flow diagrams and models, mathematical calculators like "Birth fraction" and "Death fraction" are called converters. Each modeling software program displays converters in different ways, using symbols or sometimes just using the name, as we have done in our diagrams in this lesson. Remind students to use connectors (arrows) to show that the converters affect the flows.

11. Now that students have constructed the basic diagram of a stock/flow population model, be sure they can explain the logic behind it. Any given stock of "Population" experiences a fraction (or percentage) of births and deaths each year. The "Births" add to the stock of "Population" and the "Deaths" subtract from it.

    If students will be building a working computer model, continue to step #12. Otherwise, begin the debriefing section, Bringing the Lesson Home.

12. A computer model needs numbers and equations in order to simulate the situation it represents and to produce results. Encourage students to work together to determine the values necessary for the model to run. In the following instructions, we refer both to "The Mammoth Game" and to a hypothetical town named "Anytown" as examples in order to have actual quantities for models. As an alternative to guiding the whole class through steps 12-16, have students use the self-paced handout, "Creating Your Own Population Model," p 57, helping individuals as needed.
    Begin with the stock and enter an initial value. To model "The Mammoth Game," the initial value is 20, representing the number of mammoths in the herd. In the game, dice represented mammoths. For "Anytown," begin with 1000 people.

13. Next, enter the fractions or percentages for "Birth fraction" and "Death fraction." In the first version of "The Mammoth Game," the "Birth fraction" is 1/6 and the "Death fraction" is 2/6, based on the rules set up for the dice.

    When modeling "Anytown," students can research birth and death rates for the United States, other countries, or the world as a whole. Approaching the year 2020, US birth rates were about 1.16% and death rates, 0.85%. World birth rates were about 1.85% and death rates, 0.78%.

    NOTE: Two important reminders about the values in the "Anytown" models:

    - Be sure students use birth rates per population, not per female. In our models, the

population stocks represent a total population, which is multiplied by the fraction of population that gives birth each year to calculate the birth flows.
- Convert birth rate percentages to decimals, e.g., 1.16% = 0.0116. Fractions don't require conversion but modeling programs won't automatically recognize percentages as decimals.

14. The flows will not be constant values but rather depend on other elements of the model. The "Births" flow has incoming connectors from the stock "Population" and from the "Birth fraction" converter. The number of births each year is calculated by multiplying the number of individuals in the stock by the "Birth fraction." Because the number of births will change as the stock changes, the flow of births is defined by an equation and not a constant number value. Births = Population * Birth fraction

15. The symmetry of the model should prompt students to enter the equation for the "Deaths" flow. Deaths = Population * Death fraction
The number of deaths each year is determined by the "Population" multiplied by the fraction of individuals who die.

16. Enter the time setting to run for the number of years you want to simulate. The dice version of "The Mammoth Game" simulates 20 years, so that's a good starting number. For "Anytown," a longer simulation will reflect change more clearly, so enter a time period of 50 or 100 years. Set up graphs and tables to display the results and run the model. When they have their models running, students can move on to "what if" scenarios, changing the starting values of the stock and convertors and changing the length of their simulations.

Bringing the Lesson Home

Thinking about stocks and flows in a model, what makes a population increase and what makes it decrease? When will population be stable?

Flows control the changes in a stock. Inflows increase stocks. Outflows decrease stocks. If the inflow of births is greater than the outflow of deaths, a stock of population will increase. Conversely, when the outflow of deaths exceeds the inflow of births, a population stock will decrease. When the flows are equal, a population remains the same.

How does a causal loop diagram represent population? What makes a population increase and what makes it decrease? When will population be stable?

In causal loop diagrams, births and deaths appear as competing loops. When births outnumber deaths, population grows. When deaths are greater than births, population decreases.

Is our population model complete? What might make it more accurate?

In our model, birth fractions and death fractions remain as constant fractions or percentages. In real life, rates change for both animals and humans according to the conditions encountered by the population. Animal populations in the wild will likely suffer higher rates of death during times of reduced food supply, habitat destruction, or increased hunting pressure from predators. Birth rates may rise or fall corresponding to weather conditions and other factors.

Human birth and death rates also have many influences, including average age of the population, percentage of females, economic conditions, and other variables.

What happens if a population continues to decline? Is that a realistic outcome of the model?

Running the model (or playing "The Mammoth Game") with a death rate greater than the birth rate will eventually cause the population to reach zero. Many species have gone extinct as the mammoths did. Human populations change also, but birth rates usually outpace death rates, increasing the number of people over time. Even in locations where human population declines, it rarely falls to zero. In addition, our model does not include immigration or emigration.

What happens if a population continues to grow?

A stock of population will grow if births outpace deaths but eventually it will encounter limits. Think of a pond that has fish living in it. The amount of food available and the volume of water in which they live are finite. As the fish population grows, the amount of resources available per fish goes down and will affect the birth and death rates. This natural limit is called carrying capacity. Other creatures may prey on the fish and keep their population within limits also. In natural ecosystems, the availability of food, space, and predator / prey cycles keep populations in equilibrium.

Bringing the Lesson Home
(continued)

The idea of carrying capacity for human population remains controversial. Some experts warn that we are rapidly approaching or even exceeding the ability of Earth's resources to support our current world population. Others argue that technological improvements increase our ability to provide for everyone, and sustainable practices will support population growth.

Here's one way to include carrying capacity in a population model (FIGURE 8).

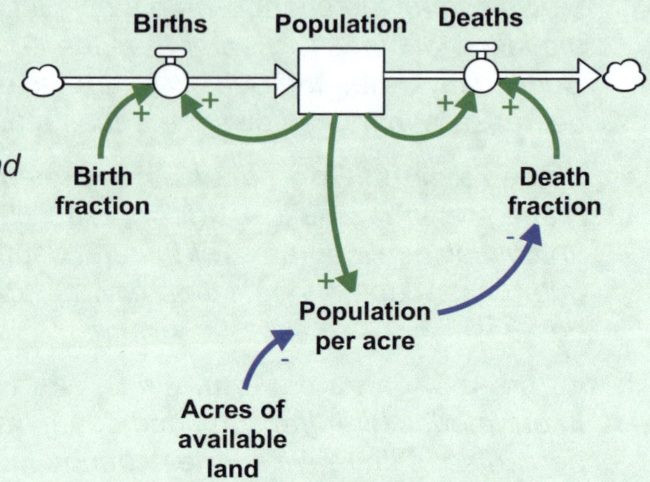

FIGURE 8

This model assumes a finite number of acres available to support the population. A large "Population" would mean there was a high "Population per acre." With fewer resources to go around, the "Death fraction" would tend to be higher. As the "Population" declined, the number of individuals per acre would be lower, and the "Death fraction" would probably decline. The converter representing "Death fraction" would change depending upon the population density. Converters that change depending on variable input have graphical functions determining their value, a more advanced modeling technique. See Appendix C for more information about modeling and modeling software.

Handout: Creating Your Own Population Model

Build the Basic Mammoth Game Model:

First, you need a "Population" with "Births" and "Deaths." You'll also need a "Birth fraction" and a "Death fraction" to show the fraction (percent) of the population that will give birth or die in any given year.

Notice that "Births" are determined by multiplying (*) the population stock by the birth fraction. "Deaths" work in a similar way. The arrows coming out from the stock to the flows show that the "Population" affects how many go in or come out.

Input the numbers into the model as indicated in the diagram. Remember, the * symbol means to multiply the variables that affect that part.

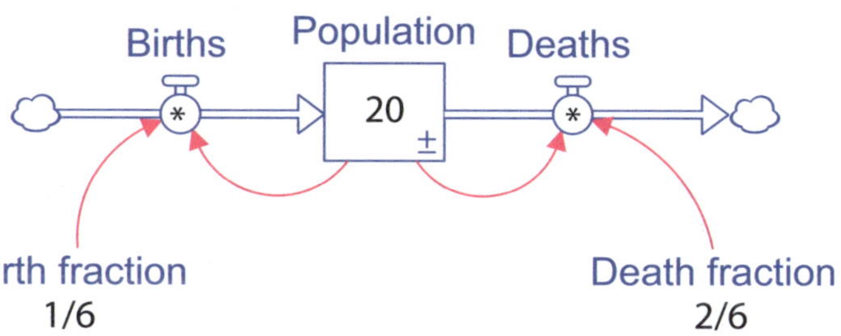

Create a graph for "Population." Run the model to see what happens. If all the initial values are correctly entered, you should see the following graph line. Notice that the line is curved. As the "Population" drops, there are fewer and fewer mammoths left that could die. If the behavior is different, recheck your connections and equations.

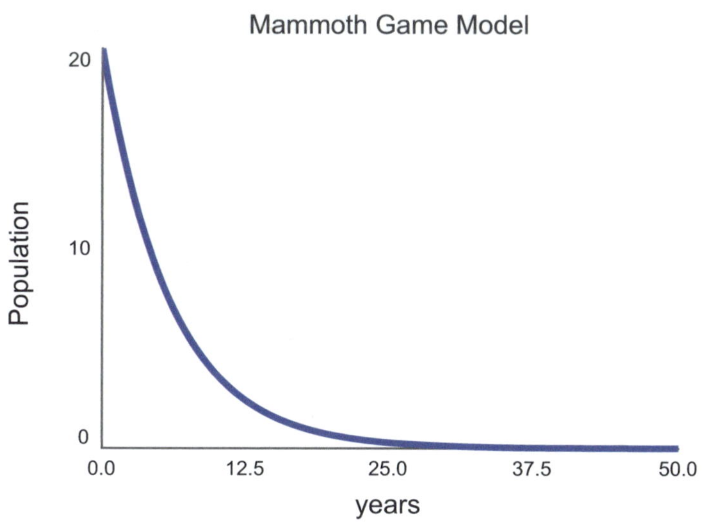

Settings

Make sure that the stock is set to allow negative values. In other words, if you are using Stella software, you must un-check the "Non-negative" box in the settings for each stock you create.

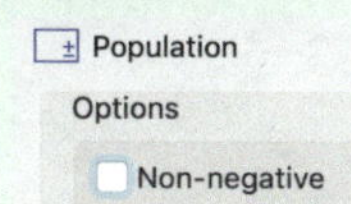

Set up the model's "Run Specs" as follows:

- Start Time = 0
- Stop Time = 50
- DT = 0.25 or 1/4
- Time Units = Years
- Integration Method = Euler

Challenge 1:

Modify the numbers in "The Mammoth Game" model to explore what would happen if the birth fraction was larger, the initial population was larger, and/or the death fraction was smaller.

Challenge 2:

Modify your model to match "Anytown USA." Approaching the year 2020, US birth rates were about 1.16% and death rates, 0.85%. World birth rates were about 1.85% and death rates, 0.78%. Remember to convert birth rate percentages to decimals, e.g., 1.16% = 0.0116.

- Population start value = 1000
- US Birth fraction = .0116
- US Death fraction = .0078

Modify the numbers for the global population rates. How is your graph similar / different?

Challenge 3:

Research populations, birth fractions, and death fractions of the world and of different countries. Make sure to use birth rates per population, not per female.

Use these values in your model to simulate the population in different countries.

Are all countries growing at the same rates? Why or why not?

Challenge 4:

Experiment with different time frames. For example, how would the results differ if you ran the model for 100 years?

When you've completed all the challenges, find a partner and repeat one of the challenges together in order to compare results. What did you learn together?

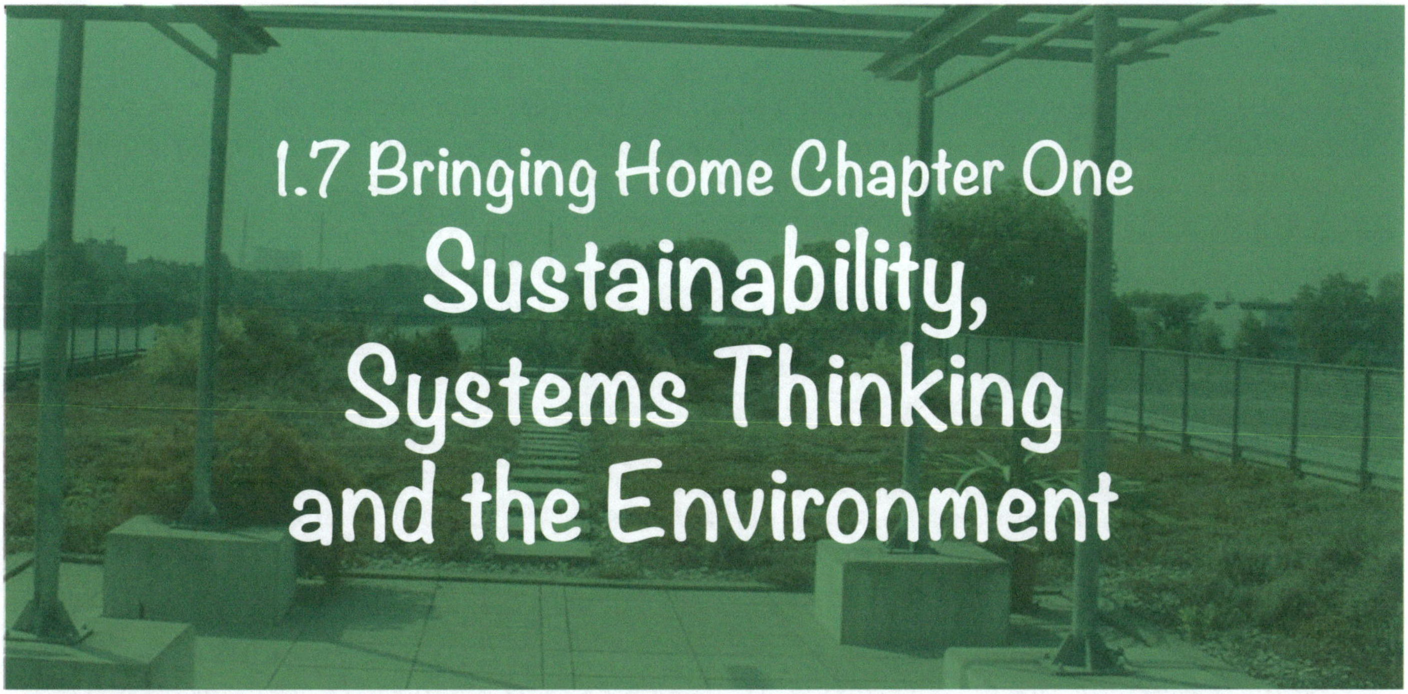

1.7 Bringing Home Chapter One
Sustainability, Systems Thinking and the Environment

How are the topics in Chapter 1 connected to sustainability?

Sustainability in modern industrial societies depends on managing growth and balancing economic prosperity and environmental systems. While each lesson addresses a specific aspect of industrial progress, a common theme is the need to maintain a healthy environment and a prosperous civilization. Cities and their transportation systems can become congested and polluted, degrading the environment and the quality of life for residents. Electricity production requires the use of natural resources and has become a major source of pollution and climate change. Managing waste responsibly can be difficult and quickly overwhelm society's capacity to control it. Environmental disasters occur with tragic frequency, warning us that we have to change our behavior and create long-term sustainable strategies. Population growth inevitably exceeds the capacity of the environment to renew essential finite resources.

How does the sustainability of environmental systems relate to other systems?

Environmental systems topics cannot be separated from other major issues affecting society. Students should be encouraged to make connections to environmental topics with three other vital systems:

- *Social justice – People of color bear a heavier brunt of the costs of environmental problems. Often prejudice and inequity lead to the worst pollution being concentrated in areas where people of color and those with less money live and work. Natural resource extraction and exploitation degrade areas where the population lacks the economic means to resist it.*
- *Economic activity – Problems with environmental systems impact business and commerce at almost every level. Corporations and individuals depend on natural resources, transportation, energy, and a clean environment. Students will be able to cite many examples of the necessity for commuting, transporting of goods, access to raw materials, ability to dispose of waste, and other essential services that depend on reliable, sustainable systems.*

- *Public health – Everyone depends on clean air and water, adequate healthy food, safe and clean shelter … what we know as necessities of life. Environmental problems can threaten our supply of these indispensible resources.*

How did systems tools help describe the importance and complexity of sustainability in environmental systems?

All the systems tools have the same goal – to describe what is changing in a system, what's important, and what needs to be done. Each tool or set of skills has a place in the repertoire of a "systems citizen."

Barry Richmond's Systems Thinking Skills help change our thinking from more simplistic and linear "cause and effect" to a deeper, more comprehensive way of framing complex problems. Systemic thinking looks at a big picture view from a variety of perspectives and emphasizes the importance of feedback. Reviewing and applying the thinking skills frequently helps students handle the difficulty of trying to understand seemingly intractable dilemmas, including how to build sustainability into a system.

Behavior over time graphs (BOTGs) show change in a system, providing a more complete picture of how a situation has gotten to its current state and what might happen moving forward. BOTGs are vital in understanding and planning sustainability.

Causal loop diagrams show feedback in systems and explain how elements in a system control each other. Most environmental systems problems contain competing loops. Humans often create reinforcing loops that continue to grow, including population, economic activity, energy production, and so on. Those loops can cause devastating effects or crash when they approach limits. Nature, however, is based on balancing loops, which are self-correcting. Controls exist in natural systems that constrain the growth of elements so the systems stays in equilibrium. Sustainability demands that we learn to live in harmony with natural systems.

Iceberg diagrams embed other systems tools and describe how events are the products of longer-term patterns, which are created by structures and mental models. Often we concentrate on events themselves and overlook their deeper causes, which can be analogous to the unseen underwater mass of an iceberg. Events are short-term in nature and reacting to them without considering long-term effects can impede progress toward sustainability.

Stock/flow (S/F) diagrams trace the movement of "stuff" in a system. If we quantify the accumulations and the rates, we can create a model to simulate their behavior. S/F thinking focuses attention on what's really happening in a system, not just what we hope will happen. Planning for sustainable systems requires that kind of operational thinking.

Chapter Two: Land
Take care of the land; it's only a small planet.

"Buy land, they are not making it anymore." Mark Twain

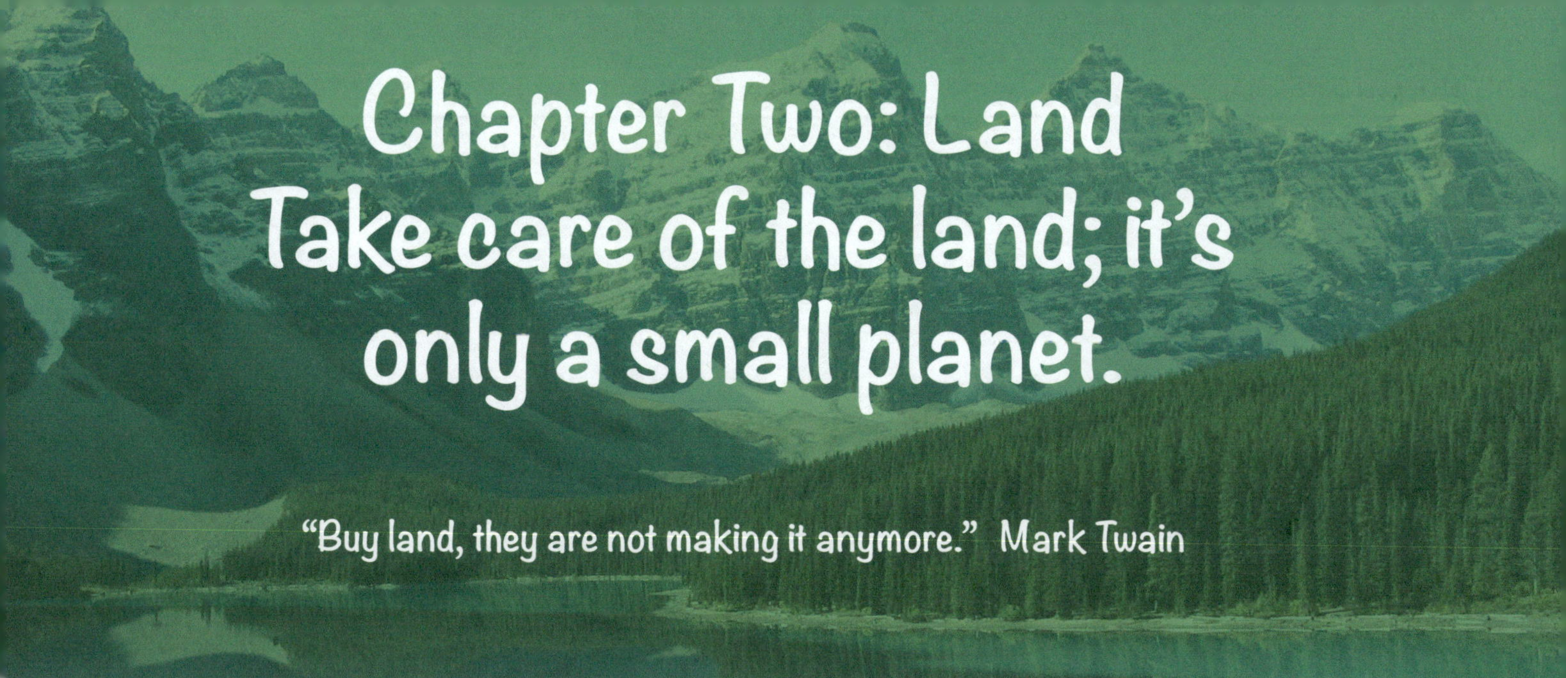

Getting started

Mark Twain's famous quote about real estate identifies land as the archetypal finite resource. Understanding land use (and misuse) is necessary in order for people to live sustainably and prosper. Students will be motivated to look forward and become responsible "systems citizens."

What will students do and learn in this chapter?

2.1 A Parable of Environmental Disaster

Study *The Lorax*, a fable about resource exploitation and degradation. Use graphs, connection circles, and causal loop diagrams to describe Dr. Seuss's warnings from *The Lorax* that resource exploitation causes environmental degradation and may not be reversible.

Understand how competing feedback loops can explain the way short-term goals can lead to long-term failure.

2.2 The Tragedy of the Commons

Roleplay stakeholders deciding how to use a tract of government-owned land designated as a national monument. Analyze systems thinking archetypes and learn how generic structures can help explain complex problems.

2.3 Protecting the Land During the Anthropocene Epoch

Graph changes that have occurred in the industrial era. Use connection circles and causal loop diagrams to describe how people around the world are dealing with the threats and challenges of pollution.

2.4 Cities and Urban Planning

Compare the effects of scale as a generic structure in different systems. Recognize the transferability of systems principles. Create feedback loops to illustrate advantages and disadvantages of cities.

2.5 Studying Rapa Nui

Investigate the history of an isolated Pacific volcanic island to understand problems of sustainability. Apply a set of systems thinking tools, including graphs, causal loop diagrams, and stock/flow diagrams and models, to understand theories about how and why events occurred as they did.

2.1 A Parable of Environmental Disaster
Lessons from *The Lorax*

Environmental Systems
How can we balance natural resource use and sustainability?

Tools

Skills

"If you cut down a forest, it doesn't matter how many sawmills you have if there are no more trees." Susan George

Overview

As long as humans have been developing technology, conflicts have arisen between using Earth's resources and preserving the natural environment. The story of *The Lorax* is a parable that points out the finite nature of resources and the dangers of pollution and overuse.

Students use systems tools to analyze *The Lorax*. They draw graphs and causal loop diagrams, preparing themselves to study "real life" examples of environmental problems.

Materials

- Copies of *The Lorax* by Dr. Seuss
- Lesson Plans on *The Lorax*, p 199
- Resource: Intro to BOTGs, p 215
- Resource: Intro to Connection Circles, p 218
- Resource: Intro to CLDs, p 220

Procedure

1. Download and review the two lesson plans available online. Decide if you want to use one or both. The first lesson, "Lessons From *The Lorax*: Using Graphs to Study Change," concentrates on behavior over time graphs (BOTGs). The second lesson, "Studying *The Lorax* with Feedback Loops," uses connection circles that lead to feedback loops.

   Based on students' prior experience, some teachers may choose to skip the first activity. However, be sure students are comfortable with how BOTGs work, especially those with multiple variables on one graph pad.

2. Have students read the story. These activities are based on the original book. Because Dr. Seuss writes in rhyme with especially fanciful language, reading aloud adds enjoyment and raises engagement in class. Some students may have seen a motion picture or cartoon version of the story, so be sure to mention that their diagrams in this activity should be based on the book.

3. Follow or adapt the guidelines in the lesson plans, which include templates and debriefing questions. The following are additional debriefing questions.

Bringing the Lesson Home

Consider this essential question: "How should human beings balance economic good with environmental harm, need with greed?" [1]

What can be learned from the *The Lorax*, in which failure to reach that balance had tragic consequences?

Although the story is a fable, parts of the Earth have been severely degraded because of industrial pollution. Students will likely bring up the necessity of regulatory oversight, as well as the need for citizens to resist pursuing a wasteful, unsustainable level of consumption.

In The Lorax, the Onceler and his family do not plant Truffula trees and soon exhaust the supply. A sustainable system needs an inflow of new materials to replace those that are consumed. In this story, and many places on Earth, that means trees.

How can humans continue to enjoy the advances of technology while avoiding the consequences of environmental degradation?

If technology got us into this mess, can it get us out? While students will be familiar with the cliché, "There is no free lunch," scientific advances in energy and other fields are promising. The three Rs of conservation can be "reduce, reuse, recycle."

Can students think of other Rs?

- *rethink*
- *repair*
- *repurpose*

Use less, use again, and use for something else: moving society away from throwing so many things away is a difficult but necessary change.

1 Book review by Beth Macy
https://www.nytimes.com/2018/03/09/books/review/behemoth-joshua-freeman.html

2.2 The Tragedy of the Commons
How Should Public Land Be Used?

Environmental Systems
How can common resources be shared and sustained?

Tools

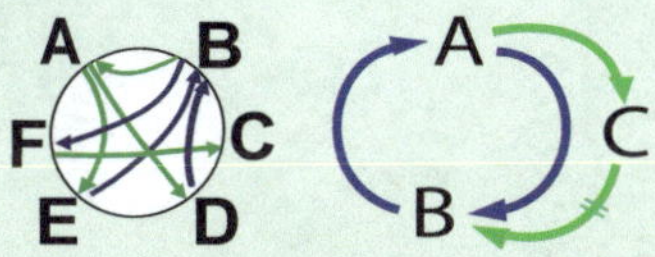

Skill

"Structures of which we are unaware hold us prisoner." Peter Senge

Overview

Land held by the government isn't always preserved as wilderness. Opinions vary concerning the goals and purposes of public land. When too many people exploit resources that are held in common, those resources may be spoiled or depleted. Environmental problems are often caused by this overuse, illustrated in a systems archetype called the "Tragedy of the Commons." Archetypes are feedback loops that describe generic behaviors - recurring actions that show up in a variety of situations.

Students tackle the problem of how a tract of land, Bears Ears National Monument, should be used. They play the roles of stakeholders and attempt to achieve a compromise solution. They identify how behavioral structures can complicate environmental decisions and learn about the "Tragedy of the Commons" (FIGURE 1) and other systems thinking archetypes.

Tragedy of the Commons

Garrett Hardin, a biology professor in California, published a now famous paper in 1968 called "The Tragedy of the Commons."

Hardin described the inherent conflict between what is optimal for society and what is optimal for each individual. When individuals use as many resources as possible to maximize their gain, the rising pressure on finite resources eventually hits a limit.

The reading level of the article can be too long and difficult for students, so the Materials list for this lesson suggests finding video and simpler explanations. Teachers may decide to provide students a summary of Hardin's article, or assign them to research it on their own and share with their classmates.

Materials

- Explanations and examples of the Tragedy of the Commons archetype, p 200
- Information about Bears Ears National Monument, p 200
- Suggested reading, p 200

- Resource: Intro to Connection Circles, p 218
- Resource: Intro to Systems Archetypes, p 228

Procedure

PART ONE – Simulating a public hearing about public land

1. In 2017-18, the US government began the process of rescinding the "national monuments" status for parts of several protected areas. One of these monuments is Bears Ears, a large tract of land in southeastern Utah. Distribute links or printed copies to students of two government documents: the fact sheet about Bears Ears issued by the US Forest Service and the Bureau of Land Management, and the later report advising a reduction in the size of the protected area.

   To set the scene for a public hearing, announce that government officials want to hear from the people who will be affected (stakeholders) by the size and use of Bears Ears National Monument. Teams of students will take on roles (assigned by chance or choice) and make their cases.

   While this activity concerns a specific situation, similar arguments and controversies over land use are ongoing throughout different regions of the US. Students will be able to research the arguments for and against reducing the monuments.

Tragedy of the Commons
(continued)

The basic principle of the Tragedy of the Commons is presented as an archetype by many systems thinkers, using causal loop diagrams as shown in FIGURE 1.

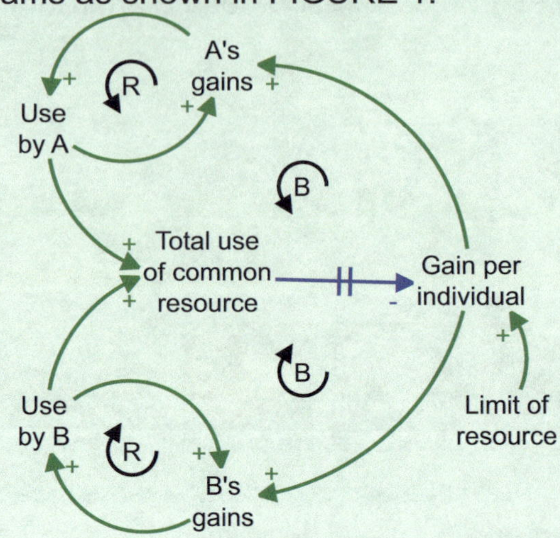

FIGURE 1

Donella Meadows wrote specifically about the Georges Bank fishing ground in the North Atlantic Ocean as a case of the Tragedy of the Commons. If you are planning to cover Chapter 3 - The Sea, hold off reading Meadows's article until then. If you are not, this short piece explains the Tragedy of the Commons well.

2. Assign teams to prepare arguments for how land in the protected area may be used. Here are ten possible teams and their interest in the land:

Stakeholder Group	Interests in land use
Ranchers	Grazing land available for cattle
Tourism operators	Off-road tours, camping, hiking
Native Americans	Protection of heritage and sacred sites
Wildlife biologists	Maintainance of habitat for biodiverse ecosystem
Energy company executives	Open land for mineral extraction
Long-term local residents	Keeping options open for possible jobs
Weekend recreation users	Keeping trails open for vehicles
Photographers, birdwatchers	Access to natural areas

Stakeholder Group	Interests in land use
Members of the media	Reporting the story and informing their audience
Government officials	Reaching an agreement and satisfying constituents

3. Each team will research the use of the Bears Ears area with special attention to their stakeholder group's point of view. Using a connection circle template, encourage them to diagram the consequences of the Bears Ears reduction plan. See FIGURES 2 and 3 for two examples, using "Size of protected area" as a common element.

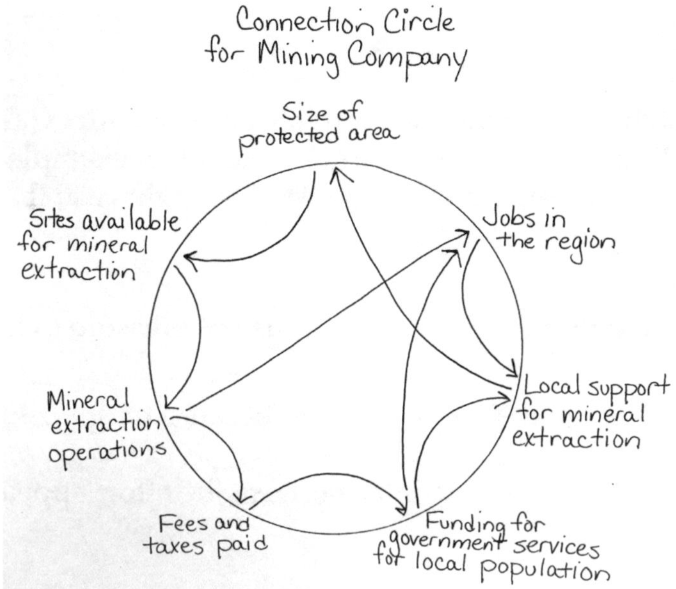

FIGURE 2

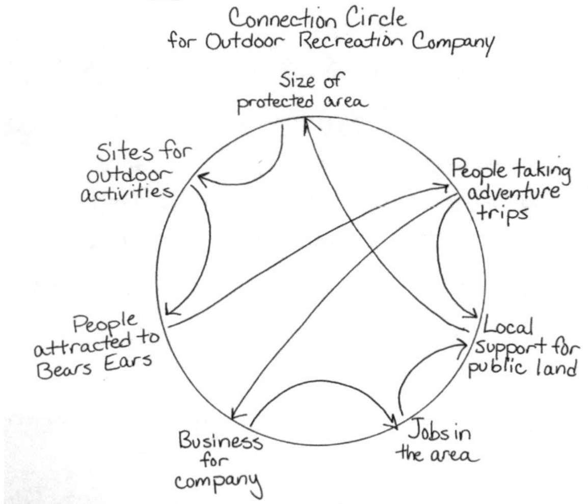

FIGURE 3

Have teams create persuasive arguments for their group's interests concerning the use of public land. Structure a class activity to allow teams to express their views. Suggested activities:

- Bring groups together for a public hearing.
- Have a "media team" interview each team, either live or on video.
- Create team posters or videos and organize a presentation for an outside audience of other students and adults.

4. How can a compromise be reached? When students leave character as stakeholders, have them propose a way to satisfy the needs and desires of all the groups. Depending on your teaching preference, this may be done in the same teams, with different peers or partners, or independently. The assignment may be to write a plan that will be presented to voters or elected officials.

When "Less is more"

This oft-quoted axiom has special meaning while using systems thinking tools. A connection circle with lots of variables can have so many connecting arrows that it looks like a plate of spaghetti! Not much insight is gained from that. Systems educators advise limiting the elements around the circle to between five and ten. After there are arrows going both in and out of each element, trace the closed loops and write them on a separate paper to see the feedback more clearly.

PART TWO – Exploring the systems archetypes

5. Give students a causal loop diagram of a Tragedy of the Commons structure. Ask how this archetype might be applied to a national park or monument.
6. Be sure they can cite other examples of a Tragedy of the Commons archetype. Here are some suggestions:
   - Fishing stocks depleted
   - Excessive logging in forests
   - Air and water pollution from overdevelopment
   - Traffic congestion
   - Environmental degradation in *The Lorax*
7. Assign one of the following archetypes to each team, using the templates in Appendix C (Intro to Systems Archetypes, p 228). Have them discuss the archetype and make up an example showing how it might apply to Bears Ears or other similar situation. What insights can they learn from the archetypes? Some examples are provided for guidance.

   Drifting Goals (also called Eroding Goals)
   A fundamental solution for reaching a goal is ignored or changed because of pressure to lower the goal.
   - reducing the amount of protection sought for a parcel of public land because of opposition pressure
   - delaying or backing off from targets like protecting sacred sites because the effort appears too ambitious and expensive to implement

   Escalation
   Competition creates an "arms race" in which two entities regard the actions of the other as a threat.
   - money spent for legal action and advertisements advocating for expanding use of public land by adversarial factions
   - resorts and recreation companies expanding and adding amenities as they compete
   - larger vacation homes built around government land
   - more powerful, noisy, and polluting recreational vehicles
   - declining willingness to compromise creating stubborn, 'dig in your heels' attitudes

   Fixes that Fail/Unintended Consequences
   An apparent solution to a problem causes unintended consequences that make the situation worse.
   - land use restrictions to protect habitats and sacred sites leading to illegal behavior and harmful consequences
   - restricting access to land causing job loss, which erodes public support for land protections

   Shifting the Burden
   A strategy that treats symptoms is used in place of a fundamental solution. Over time the fundamental solution can become ineffective and a "side effect" develops.
   - government leaving enforcement of federal regulations to local authorities
   - government regulations on companies and industries that should be policing themselves
   - park limiting visitation to prevent overcrowding, causing nearby areas to experience congestion problems

Limits to Growth (also called Limits to Success)
Growth can be rapid, building as a reinforcing loop. But limiting factors will eventually be encountered, creating a balancing loop that dampens and even reverses the success.

- private use of land, popular at first, then problems causing reversals
    - ◊ off-road vehicle use increasing, and crowds making the activity less enjoyable and more destructive.
    - ◊ welcoming mineral extraction company, jobs, and taxes are welcomed until degradation of land and pollution impact quality of life
- overpopulation causing a species to decline after it outgrows the carrying capacity of its habitat

Success to the Successful
Resources are allocated to one alternative over others, causing the favored option to succeed. The successful choice continues to receive resources and favorable treatment while the other options suffer from underinvestment.

- granting a license to use public land to a company or industry, such as mineral extraction, which grows and continues to receive support, causing alternative companies, like recreation and adventure tourism, to face obstacles gaining access to land use
- granting rights to corporations to use land because they provide jobs and pay taxes while denying use to sacred sites of Native groups, considering that purpose economically nonproductive

Bringing the Lesson Home

How can you describe the purpose of public land?

Answers will depend on different points of view. The stakeholders in the simulation in this activity all saw the land from their own perspective. Often, interested parties must compromise to reach solutions.

How could you describe the Tragedy of the Commons archetype in your own words?

When a resource is widely shared, it may be exploited by so much demand that the supply is overwhelmed. When people do not feel responsible for assets and property held in common, they may use less care to preserve them. Resources for which there is no perceived cost (like fish in the ocean) are especially susceptible to the Tragedy of the Commons.

Describe how the archetype that your team used related to Bears Ears or other public land situation.

Students can share how they matched the archetypes to environmental systems problems involving the use of public land.

What are some examples of the Tragedy of the Commons archetype besides the use of public land?

- *too many vehicles using roadways*
- *species endangered or driven extinct by hunting, like the bison or passenger pigeon*
- *trash discarded in a public or "out of the way" place*
- *water supplies being overused from a body of water or underground source*

2.3 Protecting the Land During the Anthropocene Epoch
Undoing Harm and Preserving the Environment

Environmental Systems

How can we protect and heal the Earth's land?

Tools

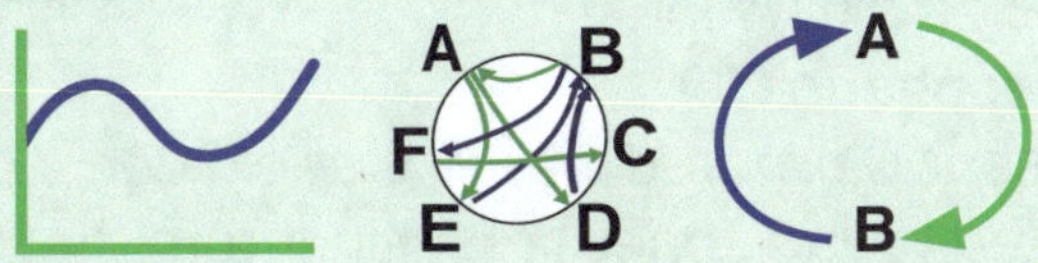

Skill

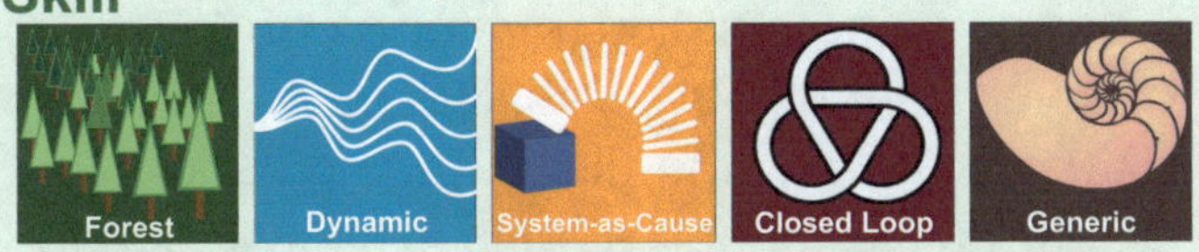

"We need the tonic of wildness ... we can never have enough of nature."
Henry David Thoreau

Overview

Many scientists believe we have begun a new geologic epoch that has been named the "Anthropocene." The pervasive influence of human activity has altered Earth so much that evidence of changes will be recorded in the layers of rock currently being formed. While geologic changes are slow by human time scales, the effects of our actions continue to accelerate and are already being felt in the 21st Century.

Students will investigate harmful consequences of human activity and analyze potential solutions to mitigate and reverse environmental damage.

Materials

- Articles and Timelines, p 200
- Handout: Thinking Skills, p 214
- Resource: Intro to BOTGs, p 215
- Resource: Intro to Connection Circles, p 218
- Resource: Intro to CLDS, p 220

Procedure

1. Introduce students to the term "Anthropocene." Distribute copies or share projected images and links from the resource list. Looking at the behavior over time graphs (BOTGs) on page 36 of the resource, entitled "When Did the Anthropocene Begin?," ask students to think of other variables to graph for the period 1800 – 2015. Encourage research to find or draw BOTGs for other data. Possibilities include

The Anthropocene Epoch

Geologists divide the history of Earth into sections of time. Dividing lines are set by changes in the formation of rock layers and formations. Until recently, geologists considered the present the Holocene Epoch, which began approximately 11,700 years ago. Since the 21st Century began, many scientists believe we have entered a new epoch called the Anthropocene, believing that human-caused changes will leave a permanent mark in Earth's rock record. Several geologic and biological systems are experiencing those changes at rates two or more times greater than the rate of natural processes.

- Human population
- Fossil fuel use
- Livestock raised
- Number of motor vehicles
- Aircraft miles traveled
- Percentage of land area covered by forest

2. Form teams of students to study the increasing impact of human activity on the ecosystems of Earth and efforts to preserve ecosystems. Use the five articles cited in this lesson plan, or have students discover topics on their own. They should derive their own feedback loops, beginning with connection circles if desired. Sample causal loop diagrams are provided for the five linked articles suggesting possible pathways and beginnings for students to follow, but encourage original work and use the examples only when guidance is necessary.

   Article: "A Vigilant Rwanda is Winning the War on Plastic Bags" (FIGURE 1)

   Students should be able to describe feedback loops by beginning with the essential problem: "plastic bags in the environment," or some similar phrase. The amount of damage plastic bags cause to the environment conflicts with their convenience and utility. As people see success cleaning up their ecosystems, a reinforcing loop builds concern for the environment, and support for a plastic ban grows.

The Anthropocene Epoch

(continued)

Geologist Marcia Bjornerud cites five of those human effects in her book, *Timefulness*.

- Erosion and sedimentation
- Sea level rise
- Ocean chemistry
- Species extinction
- Atmospheric CO_2

Professor Bjornerud warns that oxygen depleted "dead zones" resulting from agricultural runoff are expanding.

Long-lasting plastics, evidence of radioactive fallout, drastic ecosystem destruction, ... the discouraging list of destructive human effects is likely to leave its mark on the Earth, surely a call for more long-term planning and action.

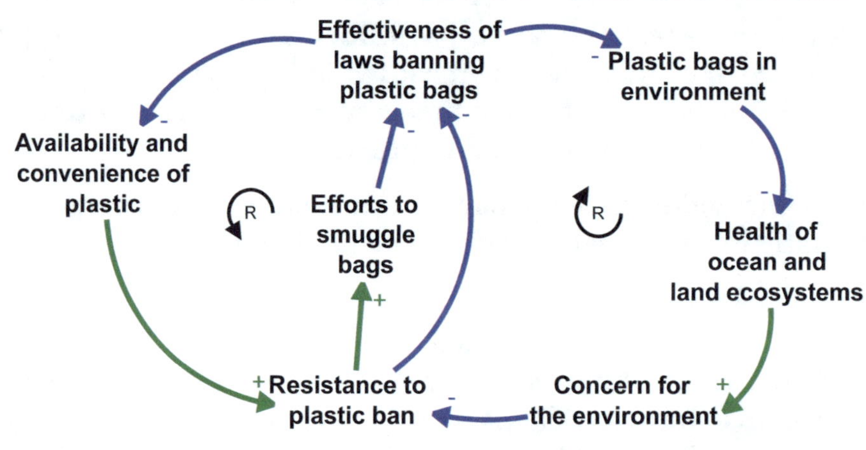

FIGURE 1

Article: "Brazil Wavers on Environment, and Wetlands Start to Wither" (FIGURES 2-4)

The tropical wetlands are in retreat, being changed by humans to accommodate farming and ranching. Brazil is a large and ecologically diverse country trying to grow its economy. Students can look at the national and regional conflicts in Brazil that are putting pressure on the nation's ability to reduce poverty and still sustain one of the world's most biodiverse regions. The first CLD considers business development versus conservation (FIGURE 2).

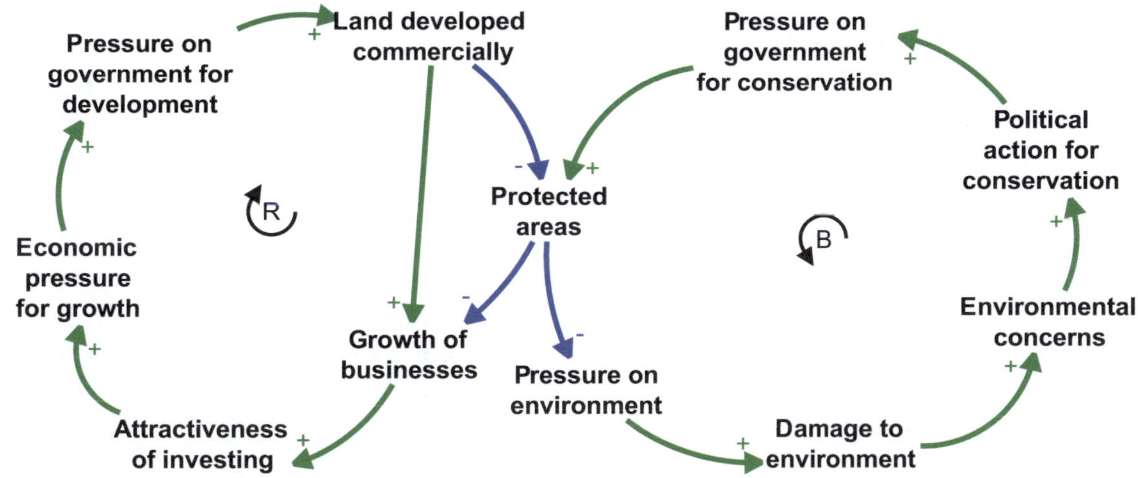

FIGURE 2

The second CLD describes the loss of undeveloped land causing long-term degradation of ecosystems. The dashed line asks whether increased concern and activism will change government policy (FIGURE 3).

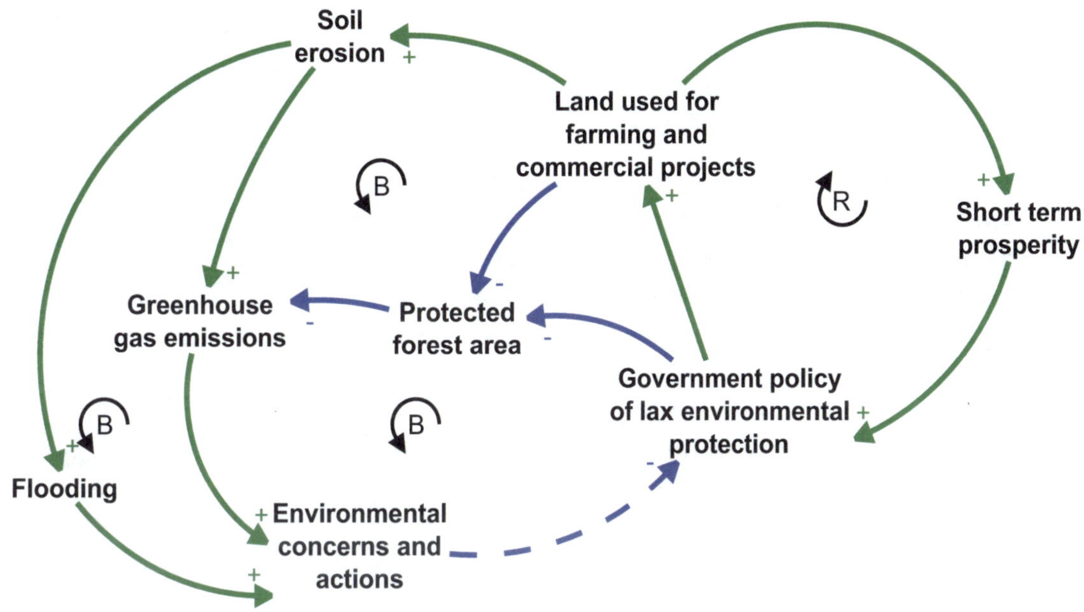

FIGURE 3.

Feedback Loops and Causal Loop Diagrams (CLDs)

Feedback loops are represented visually in causal loop diagrams (CLDs), which are a form of modeling. CLDs show how elements in a system affect each other and drive behavior. They provide a reference point for discussion and analysis, helping us ask good questions. Students' feedback loops will vary because there isn't just one correct way to draw a diagram. However, the causal arrows in the diagrams should reflect what they have learned from their research. If their CLD is based on a prediction or supposition, they should explain what led to that conclusion.

There are competing loops in most environmental systems problems. CLDs can connect multiple feedback loops into one diagram. Often, development grows and causes harm to the environment if not managed sustainably. Humans may expect economic systems to grow perpetually (labeled R for reinforcing loops), but they have limits (labeled B for balancing loops) that should send warning signs before they collapse.

The third CLD is a simplified generic view of government environmental protection threatened by economic and political pressure (FIGURE 4).

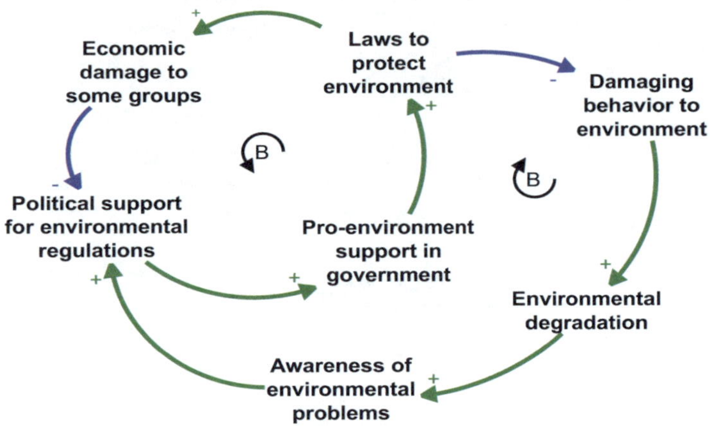

FIGURE 4

Another case of "Less is more"

Causal loop diagrams can be too complicated. Try to concentrate on the key elements in the story being told. Systems tools work best when they analyze a problem or a situation that has changed over time, rather than attempting to describe an entire complex system.

A famous quotation by Antoine de Saint-Exupery may be translated as "Perfection is finally attained not when there is no longer anything to add but when there is no longer anything to take away." Although your goal shouldn't be to build perfect models, the advice is wise – avoid loading them up with unnecessary elements.

Article: "Monsanto's Weed Killer, Dicamba, Divides Farmers" (FIGURE 5)

Throughout human history, a smaller number of farmers have been able to feed larger numbers of people. Machines, irrigation, and other innovations have increased production per person and land area, but sometimes technology can have unintended consequences. Applying chemicals to kill weeds may adversely affect the growth of beneficial plants and damage an ecosystem. Farmers can become dependent on specifically engineered seeds. A reinforcing feedback loop describes increased yields in the short term but a balancing loop warns of detrimental effects to the environment and health in the long term.

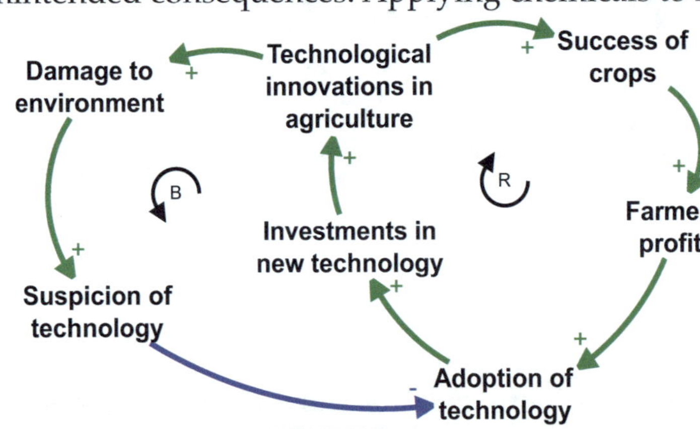

FIGURE 5

Article: "Vikings Razed the Trees. Will They Regrow?" (FIGURES 6 and 7)

Humans are incredibly adaptable, inhabiting most climate zones on Earth. Having sustained a civilization under extreme conditions for centuries, how can the population restore the environment to its prehistoric richness? Repairing the overuse of natural resources can be a challenge, even with the best intentions.

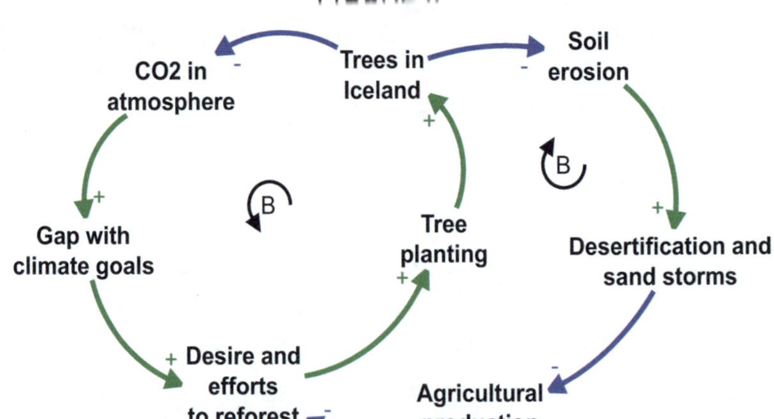

FIGURE 6

Here's a generic CLD about settlers populating an undeveloped area and causing environmental damage. Both loops are balancing loops. When the exploitation of the environment causes resources to become scarce, action must be undertaken to restore balance. Notice the double line on one of the causal arrows indicating a delay (FIGURE 7).

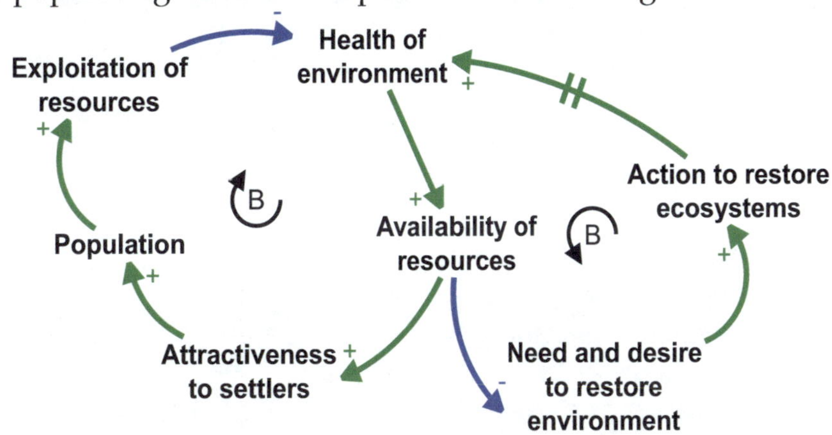

FIGURE 7

Article: "The Birds and the Bees" (FIGURE 8)

When other species suffer declines, humans need to take notice. Why are the populations of bees and hummingbirds, animals that pollinate many species of plants, decreasing? Research points to multiple factors, including the systemic presence of the family of chemical insecticides called neonicotinoids (neonics for short) in the environment. Eliminating harmful insects is an important goal, but the harm to beneficial species can be disastrous. Students can expand this CLD to include safer alternatives to neonics.

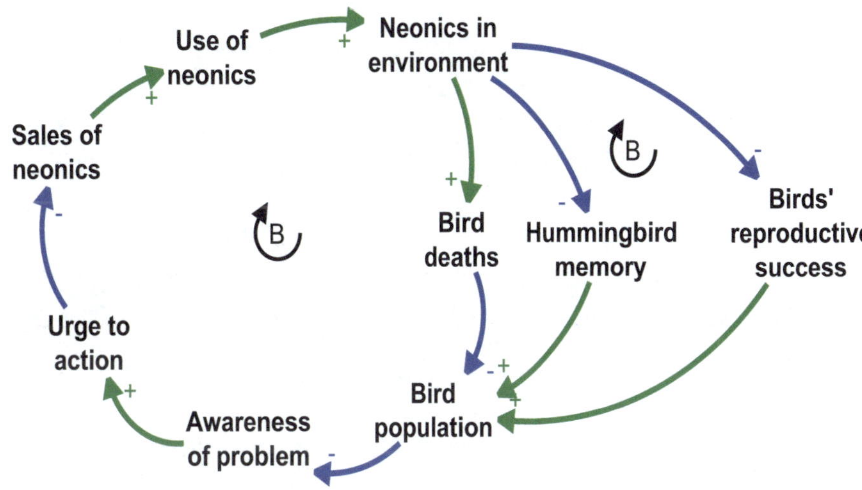

FIGURE 8

3. Compare feedback loops from the teams and look for common themes. Have students identify repeating patterns that emerge among land-based environmental systems.

Bringing the Lesson Home

How did the problem described in your article or research involve human activity?

Each team will have a different answer but common themes will likely emerge. Whether the issue is

- *plastic bags in Rwanda,*
- *degradation and loss of forest in Brazil,*
- *pesticide use and genetic engineering in the United States,*
- *efforts at reforestation in Iceland,*
- *stopping and reversing the decline of pollinators in North America,*

or another topic students investigated, humans are certainly "in the loop."

What common themes arise in the cases you studied?

Human progress toward a higher quality of life involves consuming natural resources. Discovering ways to increase comfort, safety, and overall prosperity through improved technology often takes a toll on the raw materials of Earth and damages ecosystems. Many forms of plastic have proven to be incredibly useful but most biodegrade very slowly. Forests do not grow back as fast as we can cut them down. Pesticides and herbicides allow us to choose the plants we want to cultivate, but harmful consequences follow when the chemicals remain in the ecosystem and travel where they are not intended.

Just as in the fictional story of The Lorax, real-life conflicts between human development and nature can upset the balance built into ecosystems. Damage to the environment will eventually harm the human societies that depend upon them.

What solutions can you suggest for the problems you studied?

The news is not all bad. When human behavior contributes to a problem, changing that behavior can mitigate it. Structure drives behavior – in each case studied, humans impact the environment because we use resources in a way that damages ecosystems. Changing the way we use resources may be difficult, but people can usually find a less harmful path.

- *Replace plastic bags with paper and reusable materials.*
- *Manage the resources of the land with more sustainable technology.*
- *Protect ecosystems that support people and wildlife.*
- *Reduce chemical use through integrated pest management.*
- *Support efforts to regenerate forests.*

Protecting land may cause short-term disruption and temporarily slow economic growth, but ensuring long-term sustainability and healthy ecosystems is necessary for continuing prosperity.

2.4 Cities and Urban Planning
Generic Structures and Scaling

Environmental Systems
How does the scale of cities affect their role in a sustainable future?

Tools

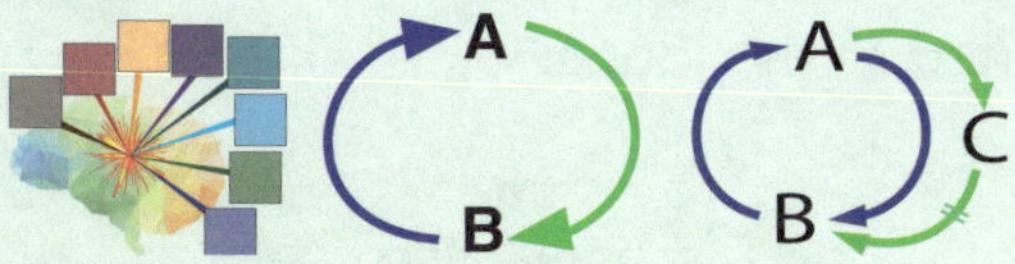

Skill

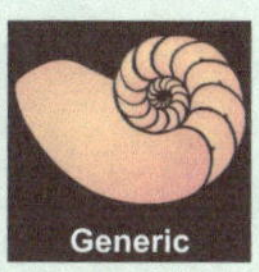

"Everything around us is scale dependent. It's woven into the fabric of the universe."
Geoffrey West

Overview

Cities share important characteristics with biological systems. Scaling up to larger size brings positive and negative consequences. While investigating the attributes of urban areas, students analyze pros and cons, drawing causal loop diagrams and citing generic structures to explain their thinking. By focusing on the effects of scale and other attributes, students will recognize generic structures that exist in complex systems.

Materials

- Articles and video, p 201
- Speech by Jay Forrester, p 201
- Handout: Attributes of Animals, p 82
- Handout: Systems Principles, p 212
- Resource: Intro to CLDs, p 220
- Resource: Intro to Archetypes, p 228

Procedure

1. Decide which of the resources about scaling to use in class. Depending on class structure, choose video and/or print material that you find appropriate for student learning modes.

2. If students did Lesson 1.1 Cities, briefly review some of the advantages and disadvantages they discussed about cities. If this is the first time they have considered cities as complex systems, have them create a chart using the one provided in that lesson – Handout: Attributes of Cities, p 17.

3. Ask students to compare the ways small animals are different from large animals. (See optional handout at the end of this lesson.) They should come up with an interesting list of characteristics. How is that comparison similar to how small towns differ from large cities? Can they describe how this mental exercise might help them think about urban planning?

   Generic structures in systems are similar patterns of behavior that occur in different contexts.

Recognizing generic structures and transferring experience and knowledge about them across disciplines adds insight to understanding complex systems.

Scaling is one important characteristic that affects complex systems as seemingly unalike as cities and animals.

4. Show video and/or assign reading material.
5. Have students work in pairs or teams to draw feedback loops that describe the advantages and disadvantages of large cities. Working first with connection circles can help find feedback loops.

   Here are some examples. Sample names have been assigned to the loops:

   "Efficiency of resources loop" (FIGURE 1)
   As cities grow, their efficiency increases. Resources are concentrated, and people are drawn to the city for economic and other opportunities. As cities scale up in size, the availability of resources can make them more attractive and productive.

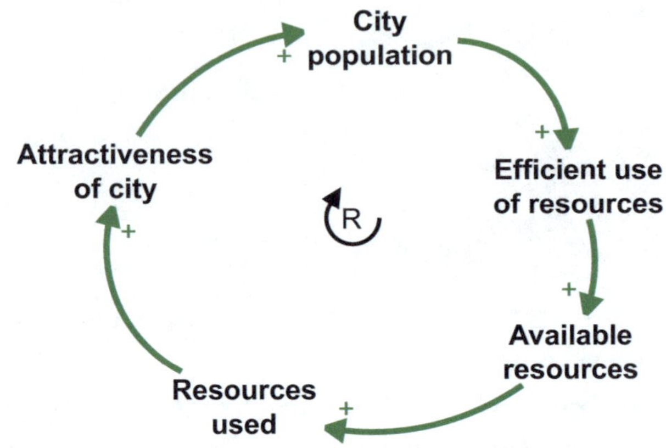

FIGURE 1

   "Social advantage loop" (FIGURE 2)
   People are also attracted to cities by all the social interaction. There is a lot happening and plenty of things to do, another advantage of scale. People can be happy and productive.

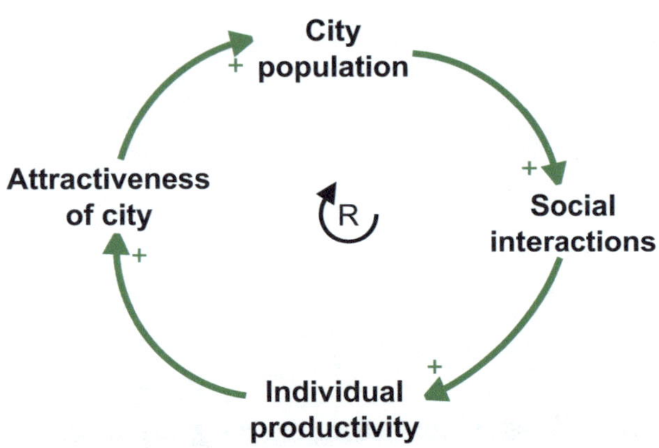

FIGURE 2

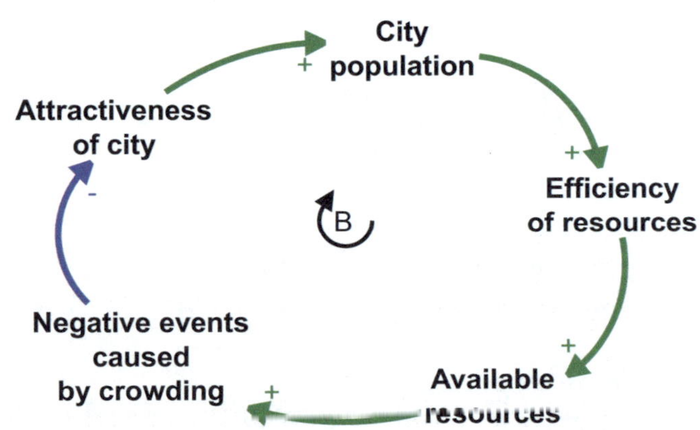

FIGURE 3

   "Problems with growth loop" (FIGURE 3)
   But as cities grow, problems become bigger also. Congestion, crime, infrastructure, services … lots of negative issues arise or get more troubling. This creates a balancing loop that acts as a limit to growth, one of the systems archetypes.

Here are the three city growth loops together in one causal loop diagram (FIGURE 4).

6. The principles of systems thinking and systems archetypes are also helpful in surfacing generic structures that affect the growth of cities. Encourage students to include some of these ideas in their urban planning analysis. Examples are cited for both of these sets in the debriefing section.

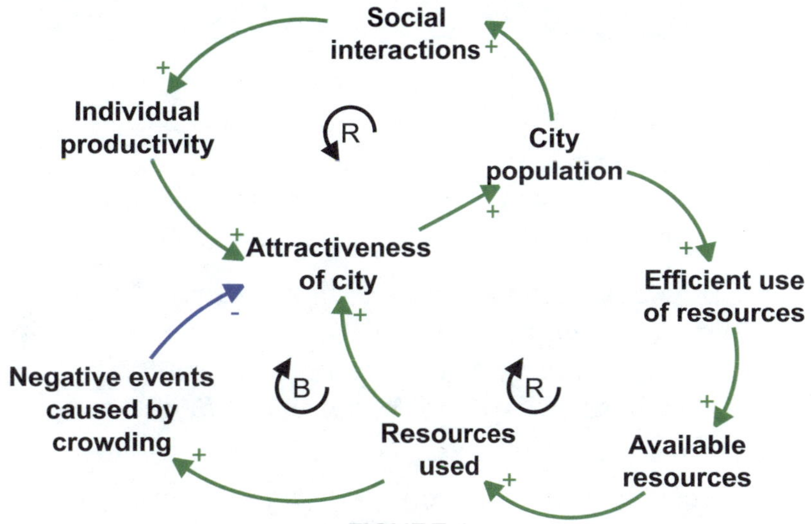

FIGURE 4

Principles of systems thinking

Professor Jay Forrester, the founder of system dynamics, frequently referred to key principles in his writing and his talks. The following four principles apply to problems that can develop as cities grow:

- What appears to be good in the short term is often bad in the long term, and vice versa.
- Cause and effect are often not closely related in time or space.
- Actions may be ineffective due to low leverage policies or incorrect application of high leverage ones.
- Almost nothing is exogenous – causes of problems are within systems (endogenous).

Systems archetypes, p 228

These systems archetypes are particularly applicable when working to balance positive and negative attributes of cities.

- Drifting Goals
- Fixes That Fail
- Shifting the Burden
- Tragedy of the Commons

Bringing the Lesson Home

How is the difference in size among animal species similar to the difference in size among cities?

In many systems, including organisms and human civilizations, increases in size bring advantages and disadvantages.

Some advantages can be summarized as "economies of scale." As more items are produced, the cost per item goes down. For example, compare building one automobile versus producing thousands on an assembly line. Economy of scale is a generic structure, a description of a condition seen in different contexts where the relationships and behaviors are similar.

Large animals can store food, regulate temperature, and otherwise use their bodies in ways that smaller creatures cannot. In cities, the concentration of population and resources creates opportunities for the benefits of civilization to flourish.

What is meant by "generic structures?" What are some examples in cities?

Generic structures describe dynamic behaviors that are similar across different systems. Relationships and causality may display parallel structures even though the contexts are unalike. Examples of systems principles that affect cities:

SYSTEMS PRINCIPLE	EFFECT ON CITIES
What appears to be good in the short term is often bad in the long term, and vice versa.	Deciding to raise taxes is unpopular but can have long-term positive impact through early childhood education, job training, infrastructure projects, and clean-up initiatives.
Cause and effect are often not closely related in time or space.	Emitting carbon dioxide and other "greenhouse gases" causes climate change years later. Other forms of pollution may also build up slowly and impact air, food, and water supplies miles away.
Actions may be ineffective due to low leverage policies or incorrect application of high leverage ones.	Lesson 1.2 Transportation examined how expanding highways can increase traffic congestion instead of relieving it.
Almost nothing is exogenous – causes of problems are within systems (endogenous).	The desirability of a city causes overcrowding that diminishes its appeal. Famous baseball player Yogi Berra, when asked if he wanted to go to a fashionable restaurant, reportedly said, "Nobody goes there anymore. It's too crowded."

Examples of systems archetypes that affect cities:

- *Drifting Goals – The goal of providing a range of low-cost housing options may be abandoned when higher cost development brings more revenue to a city and developers.*
- *Fixes That Fail – Putting people in jail for drug use and minor offenses rather than providing treatment and job training may create more serious problems later.*

Bringing the Lesson Home
(continued)

- *Shifting the Burden - City populations may have people who are experiencing hard times and are lacking sufficient food and shelter. Administrators may curtail support programs (long-term solution) because of funding problems, leaving vulnerable people to rely on charity or fend for themselves (short-term solution).*
- *Tragedy of the Commons – Tourism, especially in older cities, can overrun popular districts and hurt the quality of life for residents. The proliferation of short-term, private rentals have put pressure on many neighborhoods.*

What are some advantages to concentrating population in cities? What are some disadvantages?

Examples of positive attributes of urban centers that students may describe:

- *Abundance and proximity of economic and social opportunities*
- *Efficiency of housing, transportation, energy, food distribution, and other systems*
- *Concentration of cultural institutions, like schools, museums, performing arts groups, and other organizations*

Some disadvantages arise when a system bumps up against its limits. Cities can become overcrowded and too densely developed, causing diminished efficiency and other problems.

Handout: Animal Comparison

List how small animals differ from large animals.

Attributes of small animals	Attributes of large animals

How is this comparison of animals similar to how small towns differ from large cities?

How might this mental exercise help in thinking about urban planning?

2.5 Studying Rapa Nui
How a Civilization Declined

Environmental Systems

What happened to the civilization on Easter Island?

Tools

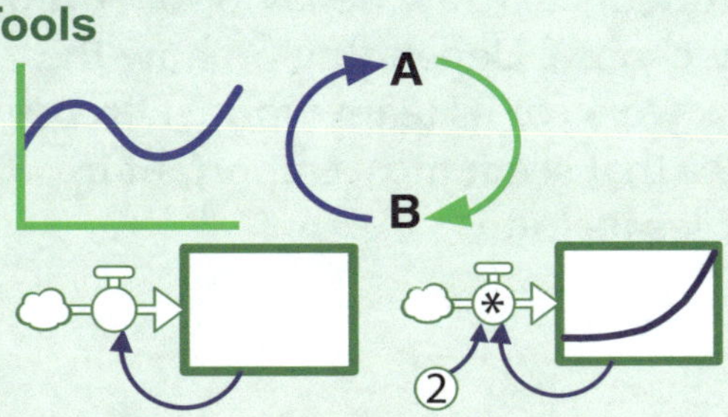

Skill

"The parallels between Easter Island and the whole modern world are chillingly obvious."

Jared Diamond, from the book *Collapse*

Overview

Rapa Nui, or Easter Island, is a remote island in the Pacific Ocean far off the coast of Chile. There was a thriving society there centuries ago, and the inhabitants built giant stone statues called moai. Eventually, the society suffered serious decline due to a variety of problems. Students will research the history of Rapa Nui and draw causal loop diagrams about its decline. They can go on to build a computer model to gain more insight into the reasons for the collapse.

About the Name

Dutch explorers sighted Rapa Nui on Easter in 1722, prompting the name Easter Island. Its name Rapa Nui comes from the local language. The Chilean government also refers to it as Isla de Pascua. In these activity plans, we refer to the island usually as Rapa Nui and occasionally as Easter Island.

Materials

- Articles and videos about Rapa Nui, p 201
- Devices with access to SD Modeling Software (optional)
- Modeling Lesson Plans, p 202
- Resource: Intro to BOTGs, p 215
- Resource: Intro to CLDs, p 220
- Resource: Intro to Stock/Flow Diagrams, p 223
- Resource: Intro to Systems Archetypes, p 228
- Resource: Intro to System Dynamics Computer Models (optional), p 224
- Resource: Intro to SD Modeling Software for drawing diagrams (optional), p 238

Procedure

1. Using a projection screen or individual computers, have students locate Rapa Nui. Either in teams or as a whole class, have them identify the island's geographical characteristics, especially its isolation and size.
2. Review the systems archetype called the "Tragedy of the Commons." Ask students to keep it in mind as they research the story of Rapa Nui. If students are unfamiliar with the archetype, please see "Intro to Systems Archetypes" in Appendix C and Steps 6 and 7 in Lesson 2.2.
3. Show video(s) about Rapa Nui and assign reading. Ask students to draw behavior over time graphs (BOTGs) about Rapa Nui with three variables they choose. Depending on how the class is structured, assign as homework, independent class work, or as team work. Discuss the graphs as a whole class and choose 5-10 of the variables that seem most important in describing the history of Easter Island. Examples of variables include
   - Human population
   - Number of trees
   - Rat population
   - Available food
   - Food per person
   - Number of moai
   - Births
   - Deaths
4. Have students construct connection circles using the variables and find feedback loops. Encourage them to cite research that supports the causal arrows they draw in their connection circles.
5. Collect students' work to display the story of Rapa Nui using causal loop diagrams. If you choose to go on to modeling, move to step 6. Otherwise, debrief the activity with questions and guided discussion in Bringing the Lesson Home.

Optional: Stock/flow and/or Computer Modeling

6. Building models in a group is usually most effective when students can talk and share ideas as they work. Either have them share computers in pairs or encourage conversation so that they can surface their thinking while they build the model.
7. Build a model of Rapa Nui following one of the plans in the Materials list. Students can draw stock/flow diagrams of the decline of Easter Island by hand or work directly with system dynamics software. In either case, be sure they can explain the logic behind their models.

Another example of "Less is more"

In stock/flow diagrams, flows can have "clouds" on one end, indicating a model boundary. Don't try to model too much. Where a particular flow comes from or ends up may be very important but can land outside of the model's parameters. In "A Skeptic's Guide to Computer Models," Professor John Sterman wrote, "The art of model building is knowing what to cut out."

8. Here is Anne LaVigne and Lees Stuntz's model from *Model Mysteries*. Their model is not specifically based on Easter Island but represents a generic description of how resources affect population.

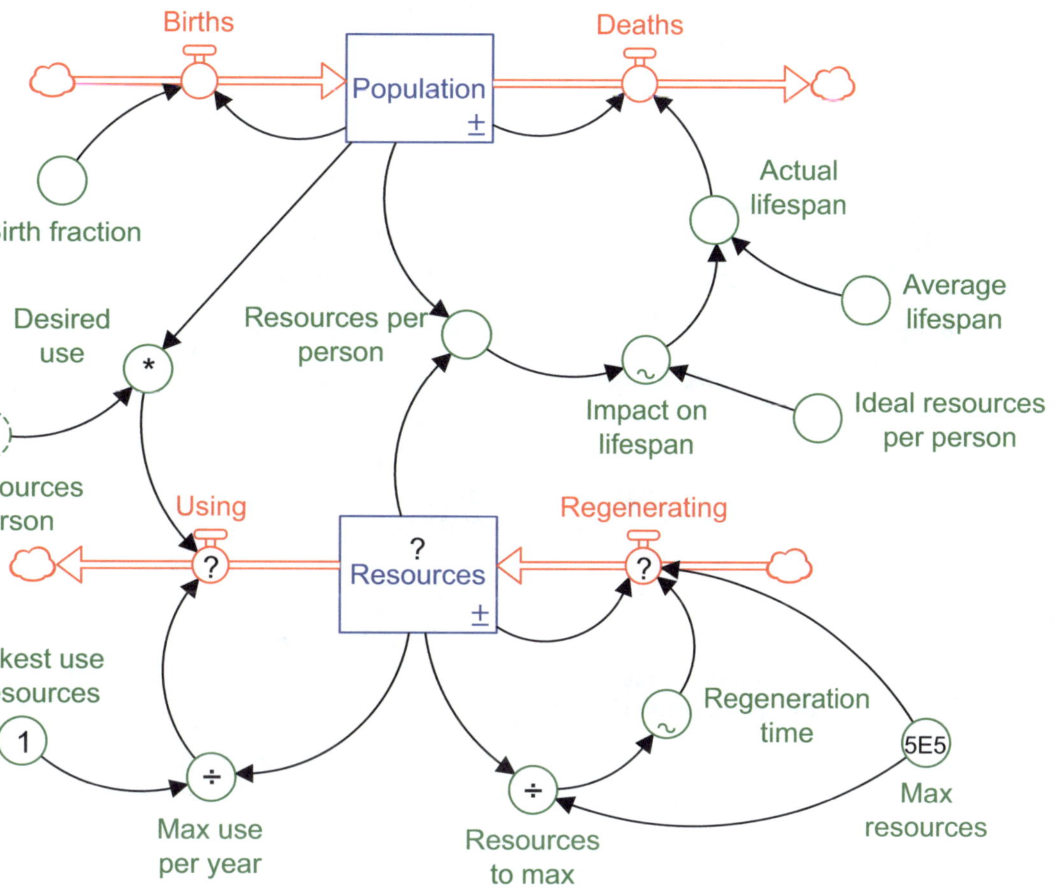

FIGURE 1

Here is a model by Diana Fisher that includes islanders captured and taken away as slaves, represented as an outflow from the population:

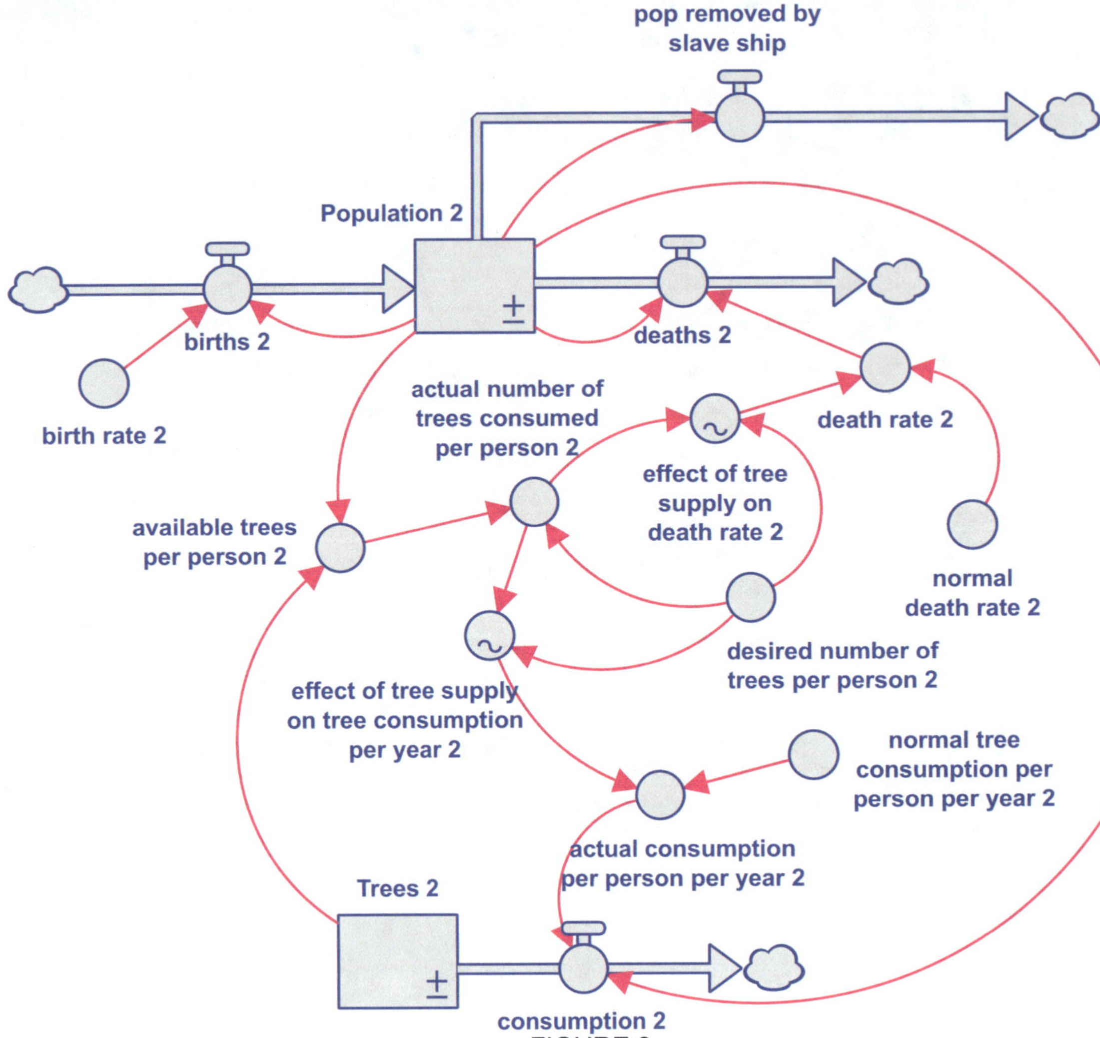

FIGURE 2

9. Students may choose to use "Food" or "Trees" as stocks. The logic in the model will be similar – resources on the island declined and the population could not be sustained. Other important elements may be added, such as a population outflow from people being captured and removed as slaves. "Rats" may be added to the model as a factor in the outflow of resources.

10. The circles in the model are converters, mathematical calculators that affect each other and stocks and flows with equations or constant values. Converter circles that contain the "~" sign are "graphical functions." That means the value of the converter is determined by the relationship between the variables connecting into it. As those variables change, the converter changes based on the ratio of those variables. For example, look at the converter in the LaVigne/Stuntz model named "Impact on lifespan." It is connected to "Resources per person" and "Ideal resources per person." When the resources per person are at or above the level of resources needed, the lifespan of people will not be adversely affected. But when that ratio falls and resources per person are below the level needed, actual lifespans will drop below average. Modelers use graphical functions to show how ratios between elements in the model cause rates of change to vary.

Bringing the Lesson Home

How did your behavior over time graphs affect your research?

The change in population over time was so large that something important must have occurred. Looking at the potential variables on the island pointed to possible reasons for the decline.

What feedback loops described what was happening?

Building and positioning the moai required the allocation of many resources. The more statues that were built, the more people competed to build them, and the more difficult it became to achieve. Conflict among groups, diminishing resources, and other factors reduced Easter Island's capacity to support its population.

The Tragedy of the Commons archetype describes part of the reason Rapa Nui failed to flourish. When resources are depleted by individuals and groups faster than they can regenerate, those resources will disappear or diminish to the point below sustainability. The island's resources could not survive the competition among groups, especially with the other factors that drove down the quality of the natural environment.

Another reinforcing loop could have driven down the tree population. When trees were cut down, there were fewer seeds produced. Fewer seeds meant fewer trees, which meant even fewer seeds the following year. Also, rats could have eaten seeds that normally would have served to regenerate the tree population. Because the rats had no natural predators on the island and their population grew, an inadequate number of seeds were left to sprout.

What lesson should be learned from studying the fate of Rapa Nui?

Multiple factors probably contributed to the decline of the island's civilization. War, slavery, overuse of resources, lack of cooperation, invasive species, and crises drove down the population and nearly eliminated the human presence there.

How could the people of Rapa Nui have avoided the downfall of their civilization?

Allocating resources poorly on a remote island combined with frequent internecine strife forms a recipe for very hard times. Leadership must maintain order and provide conditions that advance the common good as opposed to creating a climate of violence and destructive competition.

Is the Rapa Nui story unique? What causes civilizations to rise and fall?

This is a very big question, and a good one to start students thinking about sustainability. Throughout history, civilizations have come and gone. Some attain dominance over vast regions but are unable to maintain their hold. Technology, access to natural resources, climate, military strategy, trade, ideology … the list of attributes that contribute to success or failure are many and have been the subject of scholarly research about as long as we have recorded history.

Systems thinking promotes interdisciplinary study. Students who question why and how some civilizations flourish and others fade will have a rich field of study ahead, branching into nearly all academic areas.

Bringing the Lesson Home
(continued)

Additional debriefing questions for students who built models are included in the modeling activity lesson plans. Here is one more:

How did your model help you investigate the fall of the Rapa Nui civilization?

When telling a story with stocks and flows, we have to describe how things actually work. To get started, modelers can use data that are known or estimated. Then variables can be changed to try out "what if" questions with a variety of scenarios.

Models can also be expanded after the basic structure is tested. That allows modelers to test hypotheses such as the effect of rats eating seeds upon tree population decline, or the possibility of disease being introduced by European explorers.

2.6 Bringing Home Chapter Two

Land

Take care of the land; it's only a small planet.

What is meant by the "Tragedy of the Commons" and how does it affect environmental systems on land?

When heavy use of a resource held in common depletes that resource, it is called the "Tragedy of the Commons." Resources considered public and owned by public citizens are often overused. No individual or corporation has responsibility for maintaining and protecting the common resource, and therefore competition often drives it toward destruction.

The Lorax, though fictional, describes perfectly what can happen when the drive to maximize profit leads to destruction of an ecosystem. Students will be able to cite many examples of land systems spoiled or endangered by industrial pollution and sprawling development.

Can we repair environmental systems that have been damaged? What needs to be done?

Students will have researched policies and technologies that have potential to reverse environmental damage. Change can be difficult and in many cases will require sacrifice over a fairly long period of time. But progress is possible and some policies take advantage of "low hanging fruit," i.e., relatively easy-to-implement initiatives that have a positive environmental impact without affecting people's lifestyles. For example, some locations have limited or banned the use of plastic bags. Strong enthusiastic support is widespread for maintaining open space for conservation and recreation. While problems remain concerning recycling, many consumers have shifted their habits and willingly separate their waste for collection.

For longer-term, more difficult strategies, students need to keep the idea of successive approximation in mind. Even if a solution is not achieved quickly, moving in the right direction over time will result in significant improvement. A Chinese proverb says "The best time to plant a tree was twenty years ago. The second best time is now." Acting now is the best way to work for success in the future.

How does competition between human systems and natural systems create pressure on ecosystems?

Much harm is done to land-based environmental systems by people exploiting natural resources and allowing them to be degraded by pollution. Human development often results in reinforcing loops - wealth is created by extracting natural resources and that wealth is invested in more resource extraction (FIGURE 1).

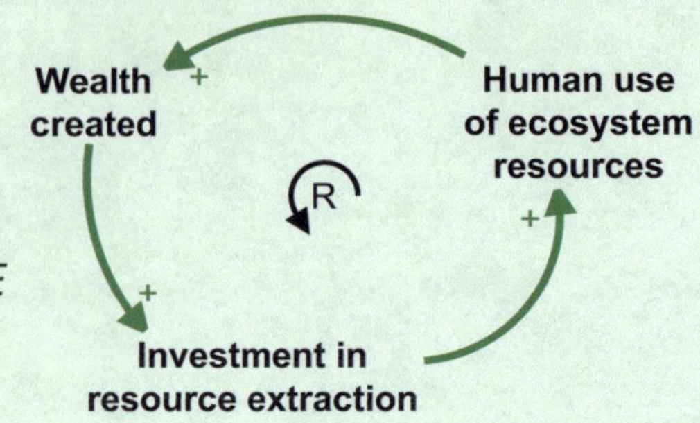

FIGURE 1

Conversely, natural systems seek equilibrium through balancing loops – population sizes are limited by the carrying capacity of ecosystems (FIGURE 2).

When the two loops compete, human reduction or destruction of habitat can overwhelm and crash the ecosystem's ability to thrive and support dependent populations. In this causal loop diagram, the element "use of ecosystem resources" represents use by humans and other organisms (FIGURE 3).

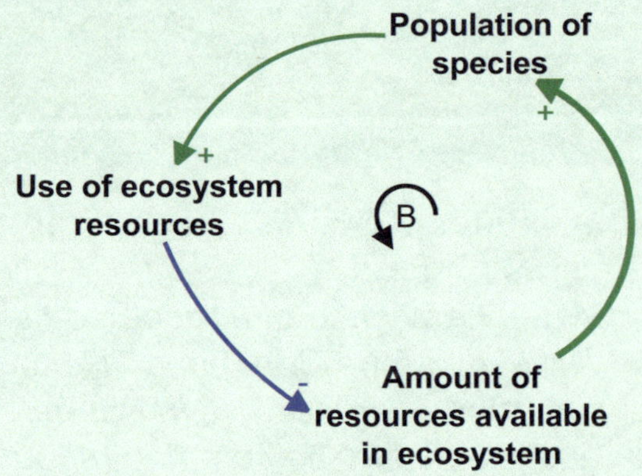

FIGURE 2

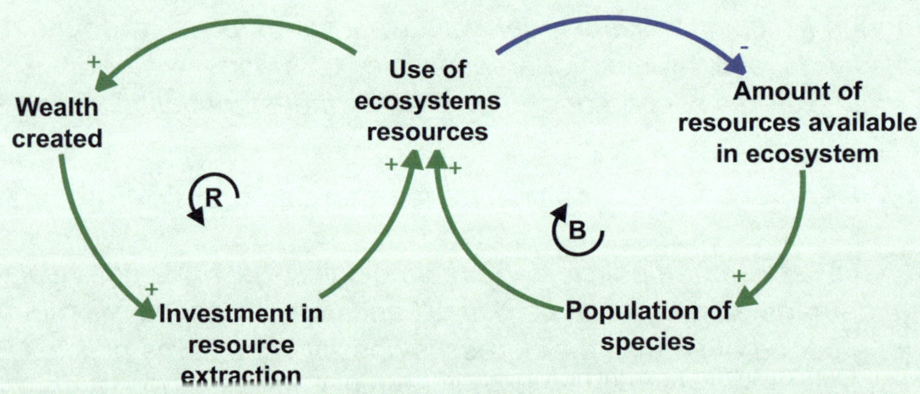

FIGURE 3

How do land-based environmental systems interact with social justice, public health, and economic prosperity?

Students will recognize the connections among the major systems in each of the lessons and generalize for most environmental issues. The burdens of pollution weigh heaviest on the poorest members of society, who suffer frequently from health crises. People with greater wealth can live away from the sources of pollution and still benefit economically from the industrialization causing the damage. Economically advantaged populations tend to consume more resources and produce more waste and pollution. While environmental problems impact the whole world, poorer people are first to bear the brunt of their effects.

Chapter Three: Air
Earth's Atmosphere is a Tiny Ribbon Around Our Planet.

"All things share the same breath - the beast, the tree, the man. The air shares its spirit with all the life it supports." Chief Seattle

Getting started

When the 21st Century began, scientific evidence and consensus had already connected human activities to Earth's changing climate. Harmful practices continued to increase, which included mushrooming carbon emissions from fossil fuel combustion, clearing of large areas of forests, wide-spread pollution of water and land resources, and other consequences of a growing population with expanding needs and desires. This chapter asks questions about what is going on and what might happen in the future. Can we have access to technology and sustain a healthy environment?

System dynamics modeling is a powerful tool to address questions and problems. Modelers can change variables in a simulation, allowing them to make predictions based on quantitative data and to make policy recommendations. Students will investigate and model the atmosphere and be better prepared to make decisions as responsible citizens.

What will students do and learn in this chapter?

3.1 Protecting the Atmosphere

Research air pollution and efforts to reverse it. Use causal loop diagrams to explain current conditions and what might be done to improve air quality.

3.2 Climate Change

Graph changes in air temperature and other variables. Use feedback loops to show the relationship between climate change and the severity and frequency of wildfires.

3.3 Understanding Stocks and Flows

Practice stock/flow thinking through activities and challenges. Explain how reducing greenhouse gas emissions may not reverse climate change without a significant delay.

3.4 The Carbon Cycle

Follow the path of carbon, diagramming it with stocks and flows and graphing its concentration in the atmosphere.

3.5 The Greenhouse Effect

Build mini-environments to test the heat-trapping properties of various materials. Consider how feedback effects push temperatures on Earth higher.

3.6 Global Warming

Study the effects humans have had on average global temperatures. Present findings using systems tools like behavior over time graphs, causal loop diagrams, and iceberg diagrams.

3.7 Examining Climate Models

Investigate the most recent climate science. Participate in a "world summit," striving to reach an agreement that will prevent global temperatures from reaching levels that will have catastrophic consequences. Use climate models to forecast future conditions and suggest strategies moving forward.

3.1 Protecting the Atmosphere Breathing Fresh Air

Environmental Systems
How does human behavior endanger the Earth's atmosphere?

Tools

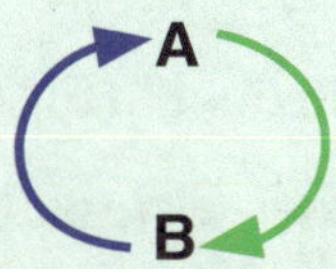

Skills

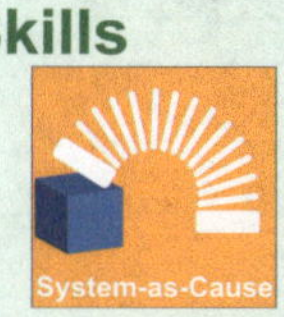

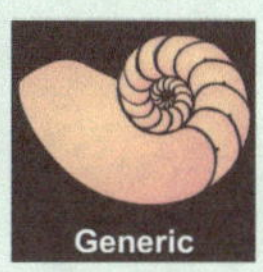

"We pay for power plant pollution through higher health costs." Sheldon Whitehouse

Overview

Air pollution remains a persistent problem in the United States and other nations. It is a global problem, not a local one confined to cities and industrial areas. Like many countries, the US has regulations that limit emissions from industrial facilities, but enforcement is inconsistent.

Students research human threats to Earth's atmosphere and explain them using causal loop diagrams.

Materials

- Background information, p 202
- Articles about air pollution, p 202
- Handout: Good News/Bad News, p 98
- Handout: Connection Circle Template (optional), p 219
- Resource: Intro to Connection Circles, p 218
- Resource: Intro to CLDs, p 220

Procedure

1. Show the video, "How Does Air Pollution Affect Health?" Ask teams to list "good news" and "bad news" reported in the video.

   Examples:

Good news about air pollution	Bad news about air pollution
Pollution is decreasing in US and some developed countries.	Air quality remains poor in many developing countries
Health improves with cleaner air. Long-term studies have provided scientists with favorable data.	People breathing polluted air are more likely to contract disease and live shorter lives.

Good news about air pollution	Bad news about air pollution
Controls on sources of emissions have improved air quality in many nations. Technology continues to improve.	Combustion is the source of much air pollution, as well as increased global warming.

2. Assign suggested articles to each team, or generate topics and have them do research to investigate air pollution. Students should derive their own causal loop diagrams, beginning with connection circles if desired. Sample causal loops are provided below for the four linked articles suggesting possible pathways for students to follow, but encourage original work and use the examples below as a starting point when guidance is necessary.

   Article: "Smothered by Smog, Polish Cities Rank Among Europe's Dirtiest" (FIGURES 1-2)

   The first diagram shows a connection circle. The second diagram shows a causal loop diagram (CLD) derived from the connection circle. Students can use connection circles as a brainstorming tool and a step toward drawing CLDs.

FIGURE 1

The competing loops in the CLD (FIGURE 2) demonstrate the conflicts in most industrial and developing economies. Fossil fuel resources drive businesses and provide energy for homes. But the pollution from emissions degrades air quality, leading to health problems and climate change. The government has responsibility to protect citizens' health and also to grow the economy while modernizing energy production. Notice the two reinforcing feedback loops connected to "Public support for clean energy." "Educating young people about clean energy" builds the population of people advocating for clean energy initiatives. Also, "Government programs" increase support for further initiatives. "Furnace replacements," for example, will make people likely to support other modernization programs in the future.

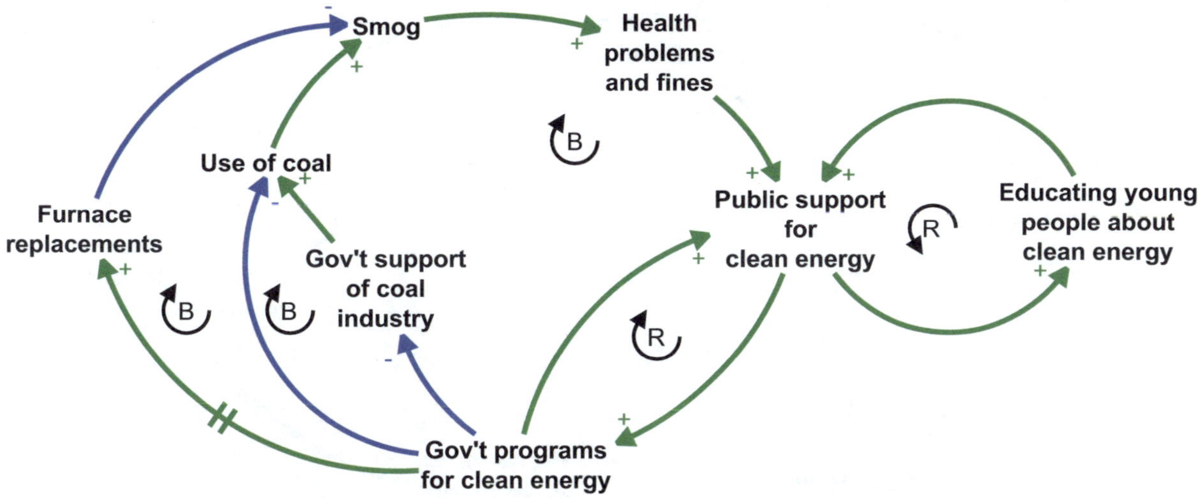

FIGURE 2

The two parallel lines on one of the causal connector arrows are a symbol for a delay in the system. Encourage students to use the double lines symbol to identify delays in the CLDs they draw.

Article: "Ship Exhaust Makes Oceanic Thunderstorms More Intense" (FIGURE 3)

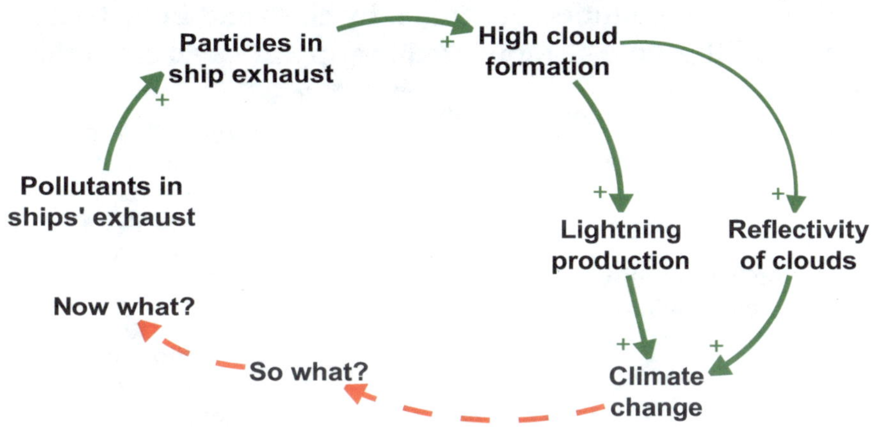

FIGURE 3

In this example, the CLD is not closed. The dashes on connecting arrows ask questions. What will be the effect of climate change on people's willingness to limit pollution? Have the students who read this article speculate on the significance of the research. What is important about strong evidence that human activity affects weather and climate? What does it mean for future policy decisions?

Article: "New Delhi, Its Air Toxic, Declares a Health Emergency and Closes Schools" (FIGURE 4)

Air pollution is causing significant problems in New Delhi and other cities in India. Can the government act strongly to curb emissions that foul the air and impact health, daily life, and economic activity? Will citizens insist on changes to clean the air or accept the air pollution as an unfortunate side effect of 21st Century life? Pressure on the government to act to limit air pollution would create the balancing loops in this CLD (FIGURE 4).

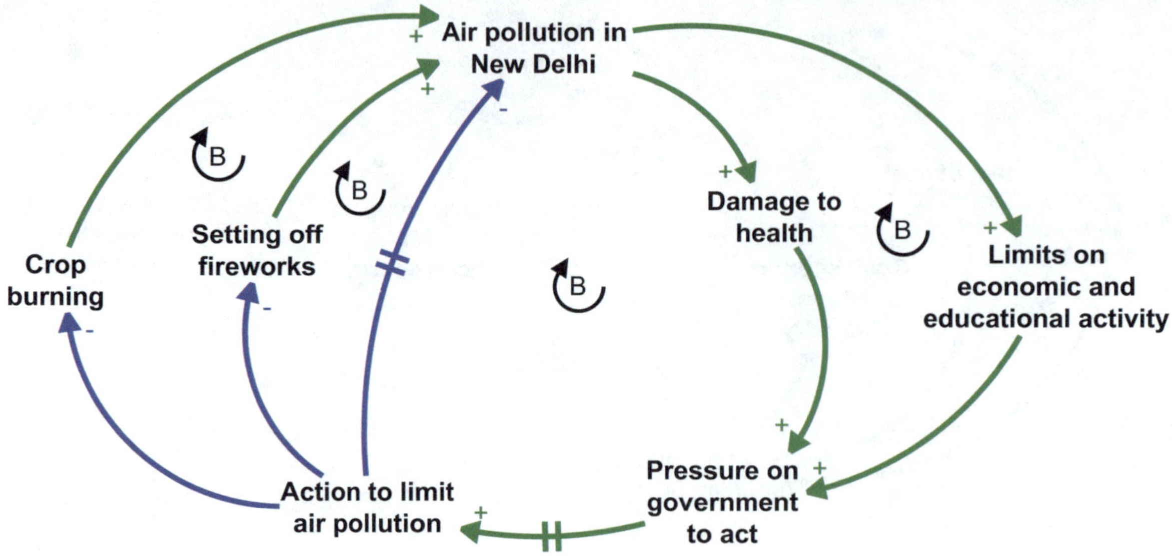

FIGURE 4

Article: "Popular National Parks Experience as Much Ozone Pollution as Major Cities" (FIGURE 5)

National parks serve as oases for people to experience natural beauty and serenity, but research has revealed that air pollution generated by cities and industrial areas impacts air quality in some parks. Will people be motivated to support clean air initiatives as a result?

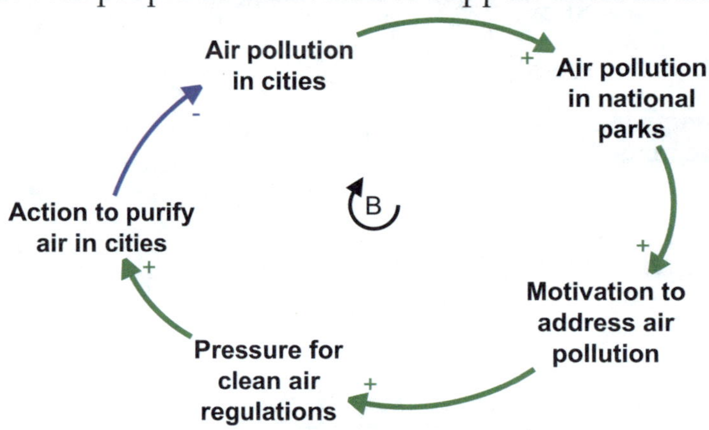

FIGURE 5

3. Share the diagrams with the whole group, either by poster session or presentations.
4. Have each team look for common structures and themes among the articles based on the CLDs.

Bringing the Lesson Home

What common themes emerge from your research and reading?

Air pollution impacts people in many ways all around the world. Health suffers and businesses cannot function at full capacity. Overall quality of life declines when the air is dangerous to breathe.

Governments need to keep their citizens safe while maintaining a strong economic climate that promotes prosperity. The articles describe how the business and environmental feedback loops have begun to affect each other, causing people's quality of life to decline.

What can be done?

Governments must strike a balance between economic growth and environmental protection. Systems thinking teaches us that short-term gains can be detrimental in the long-term, and that short-term sacrifice may be necessary for long-term success. Limiting air pollution requires shifting away from using large quantities of carbon-based fuels, which causes disruption during the transition. But if we continue on the path of high carbon emissions, we risk destroying the ecosystems that provide our clean air.

How do causal loop diagrams (CLDs) help us understand the causes of air pollution and potential strategies for change?

CLDs trace causality in a system and describe connections among behaviors. Even when the causes of air pollution are clear, the solutions involve changes that can be difficult to implement and unpopular with many people whose businesses and jobs may be hurt in the short-term. Usually, people need to reach a decision point where they will influence the government enough to act to protect the environment.

The elements of a CLD affect every other element in the feedback loop. Reversing the direction of change of one element can change the direction of the system. For example, when the effectiveness of government enforcement of clean air regulations increases, air quality improves. When enforcement of regulations decreases, air quality declines. Tracing causal arrows to determine how to increase government enforcement of regulations can result in strategies that move toward long-term solutions.

Handout: Good News/Bad News About Air Pollution

Listen and watch the video carefully to discover both good and bad news about air pollution. Write your discoveries:

Good news about air pollution	Bad news about air pollution

What questions do you have about pollution? Which question are you most interested in exploring?

3.2 Climate Change
Rising Temperatures and Larger Fires

Environmental Systems
What do we know about our changing climate?

Tools

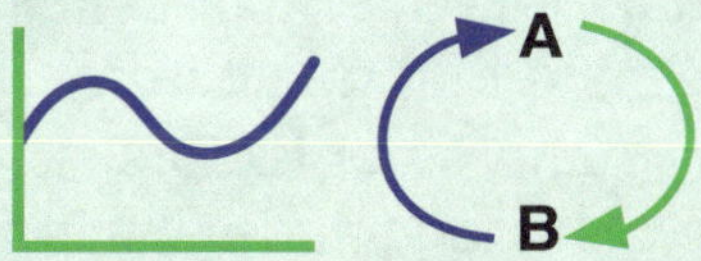

Skill

"Everybody is entitled to his own views. Everybody is not entitled to his own facts."
James Schlesinger

Overview

Climate change is a "big idea," a topic frequently in the news and under discussion in many different contexts. Even though scientific consensus strongly supports human behavior as the primary cause of climate change, significant segments of the population refuse to acknowledge the need for fundamental change.

Students brainstorm their current thinking about climate change. Then they research the topic, comparing their previous mental models to the conclusions they draw afterward. They compile a set of behavior over time graphs that illustrate what climate changes are occurring. Next, they investigate the connection between increasingly strong wildfires and climate change, illustrating their conclusions with causal loop diagrams.

Scheduling this lesson

This activity will take more than one class period, especially if you do Parts One and Two. Because all students work at different paces, exercise flexibility when scheduling. Vary the amount of work done in teams versus whole class instruction and discussion as appropriate for your students.

Materials

- Markers and large poster paper
- Background information on climate change, p 202
- Articles about wildfires, p 203
- Handout: Behavior Over Time (optional), p 217
- Handout: BOTG Template (optional), p 216
- Resource: Intro to BOTGs, p 215
- Resource: Intro to CLDs, p 220
- Resource: Intro to SD Modeling Software for drawing diagrams (optional), p 238

Procedure

Part One

1. If students are unfamiliar with behavior over time graphs (BOTGs), use the resources in Appendix C to introduce them to this tool.

2. Activate student interest and engagement by having them brainstorm in teams about the topic of climate change. Have each team divide a large poster-sized piece of paper into four quadrants. Label the quadrants,

   a. What?
   b. How?
   c. Why?
   d. So What? and Now What?

> **What is the difference between global warming and climate change?**
> Global warming refers to the measurable increase of the Earth's average temperature over time. Climate change is a term describing changes to Earth's climate that result from the rising temperatures and other factors. Indications of climate change include shifting precipitation patterns, severity of weather events, and other changes.

3. Give students the following four prompts and ask them to fill in their posters using their current knowledge and their opinions, i.e., mental models, about the topic of climate change. Use the optional handout, "Behavior Over Time" if desired. They may write statements or questions in any of the poster's quadrants. These prompts are helpful when confronting any complex problem.

   a. What is changing? Identify the variables in the climate situation.
   b. How is it changing? Draw behavior over time graphs to describe the trend or trajectory.
   c. Why is it changing? Think about the causes and structures driving the behavior.
   d. So what is the significance? Now what? Consider what should be done based on the information and insight gained.

4. After teams have filled in their posters, have them highlight or mark statements and questions about which they want to learn more. Tell them to keep these topics in mind as they research climate change.

5. Divide the web resource links among teams or have them choose their own research sites. Ask students to compile a set of BOTGs that they find important and interesting while they watch videos, listen to audio, and read. BOTGs can be sketched rather than plotted point by point, but should be clear enough for sharing with the class. Use BOTG template if desired.

6. After teams have done their research and compiled their BOTGs, have them add or change information on their four-quadrant poster based on their work. Have them write down important information that confirms, explains, or contradicts their own ideas. Have they changed their mental models?

7. When each team shares their BOTGs with the whole class, they can use the four prompting questions listed in Steps #2 and #3 to structure their presentations.

Part Two

8. Assign each team one of the articles about wildfires or have them find one themselves. Can they relate the increasing danger of wildfires to the issue of climate change? Have them present their thinking to the rest of the class using visual tools like BOTGs and feedback loops when appropriate. Allocate sufficient time for questions.

Notes on the suggested articles and related causal loop diagrams:

Article: "Fire on the Mountain: 2 Forests Offer Clues to Yellowstone's Fate in a Warming World" (FIGURE 1)
Forests are changing, perhaps due to increasing frequency of fires.

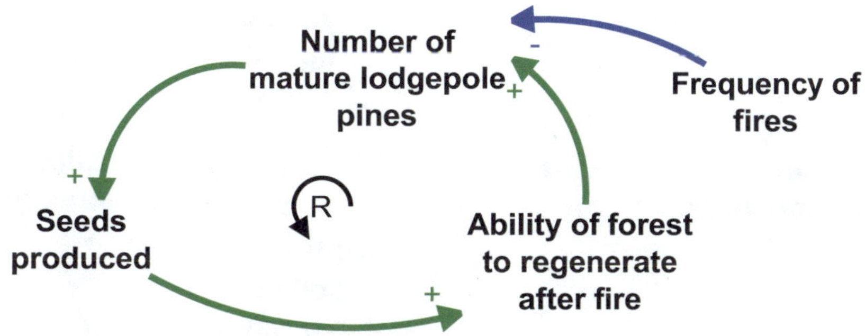

FIGURE 1

Article: "The Fire Next Time"
Text of book reviews discussed in a book review podcast.

Article: "Are American West's Wildfires Inevitable?" (FIGURE 2)
Wildfires and climate change have created a dangerous feedback loop.

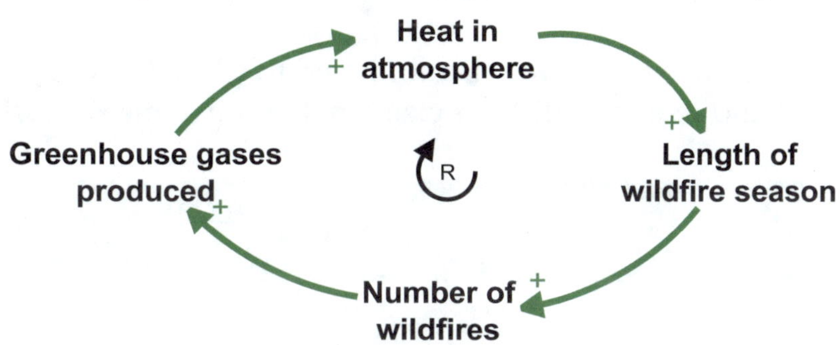

FIGURE 2

Article: "An Uncertain Future" (FIGURE 3)
Climate change is affecting national parks in the US, threatening ancient dwellings.

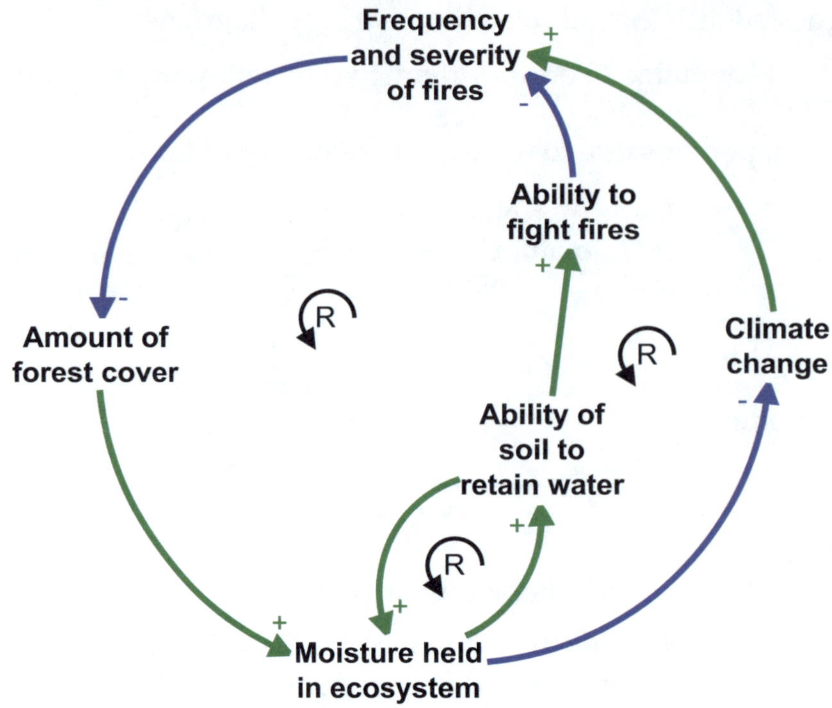

FIGURE 3

Article: "Australia's Wildfires Have Now Been Linked to Climate Change"
Warmer temperatures and less rainfall have exacerbated the damaging wildfire season in Australia.

Bringing the Lesson Home

What variables did you identify from scientists' study of climate change?

Temperature, sea level, CO_2 and other gases in the atmosphere, and additional quantitative data are all being watched closely. Other indicators include frost dates, precipitation, strength and frequency of storms, fires, and other climate-related factors. Be sure students can summarize and describe data they researched. The exact figures are not as important as an understanding of the overall trends.

How are these variables changing?

Students will describe rising temperatures and sea level, increased CO_2 levels, and other evidence.

Why is the time scale (x-axis) important on climate BOTGs?

"Weather" describes short-term conditions while "climate" refers to long-term trends. Weather fluctuates from year to year. Data on a graph with a short time period can be indicative of unusual events and seasons. When a time horizon is longer, it allows us to "step back" and see a bigger picture. Then we can determine what direction the data are trending.

A longer time period also covers more history. Tracking CO_2 levels and other variables from pre-industrial eras gives us insight into the effects of human activities on climate.

Bringing the Lesson Home
(continued)

Why are these climate indicators changing?

Most students will cite the data that show CO_2 levels in the atmosphere rising since the 1700s. Because CO_2 levels correlate with temperature levels, fossil fuel burning by humans is believed to be causing global temperatures to rise. Rising temperatures cause other factors to change, such as sea level rise from expansion and melting ice, as well as affecting weather patterns.

What causal evidence links CO_2 to warmer global temperatures?

Since the 1700s, human activity has caused an increasing amount of CO_2 to enter the atmosphere. The per capita emissions of CO_2 have steadily risen, and the population has grown more than seven times greater. More CO_2 means more heat is trapped by the atmosphere, driving temperatures higher.

So what? Why are rising CO_2 levels and rising temperatures a problem?

Global warming can have many potential negative consequences. Here are a few examples students may mention:

- *Fresh water supplies may become unreliable as precipitation patterns change.*
- *Sea levels may rise, flooding coastal and island communities.*
- *Diseases and infestations may spread as the organisms that carry them increase their range.*
- *Ecosystems could become disrupted as species are forced to move due to climate change.*
- *Ocean chemistry might change, endangering vital food resources.*

What may be the link between climate change and forest fires?

As the climate in many areas becomes warmer and drier, forest fires have become more frequent and more destructive.

A large majority of climate scientists believe human activity is a primary cause of climate change. However, many people in the United States do not agree. What accounts for this difference?

Despite increasing consensus among scientists, public opinion has moved more slowly toward accepting the role of carbon emissions in raising global temperatures. The factors that make people skeptical are complicated:

- *Manipulating data - Earth's climate is a very complex dynamic system. Data are available across many years and most regions of the planet. Some people try to pick and choose data sets that don't reflect overall trends but only target certain areas or time periods.*
- *Lack of understanding - Climate science is complicated. In addition to rising temperatures, climate change brings many effects, including stronger storms, rising sea levels, changing patterns of precipitation, habitat change, and more.*
- *Not wanting to believe - "An Inconvenient Truth," the title of Al Gore's book and movie, implies that changing our behavior will require effort and changes in lifestyle. Continuing to burn fossil fuels at the rate we have become accustomed to will lead to potentially catastrophic changes. That's hard to accept.*

Bringing the Lesson Home
(continued)

- *Short-term versus long-term – Weather and climate differ in their time scales. A cold spell in the winter with locally heavy snowfall doesn't mean that global warming has reversed. For most regions, the average temperature over time has risen steadily, but short-term weather conditions vary and may contain wide extremes.*
- *Political considerations – Governments make policies based on rules and laws. Who makes those rules? Decision makers have difficult choices to make. Protecting the environment and providing citizens with a high standard of living may create conflicts, causing people to "choose jobs over the environment."*
- *Economic considerations – Many large corporations profit from fossil fuels. Executives of those companies often work hard to convince the public that their organization is not harming the environment.*
- *The nature of science – What we accept as true today may be disproved in the future. There is always some degree of uncertainty. Science strives to provide the clearest, most accurate explanation of the world given the data available. Students have likely studied the "scientific method" and will know that systematic testing of hypotheses advances our understanding. Adding increasing amounts of* CO_2 *to Earth's atmosphere is like an experiment on a massive scale.*

Many Americans believe modern civilization has conquered nature and we should continue with "business as usual." Even when catastrophic weather causes extreme damage and loss of life, the events are often characterized as anomalous, once-in-a-lifetime occurrences. But the frequency and intensity of climate-caused crises are adding to the majority of Americans who support action to address this global issue.

3.3 Bathtub Thinking
Department Stores and CO_2

Environmental Systems
Why are atmospheric levels of CO_2 likely to continue to rise even if we reduce emissions?

Tools

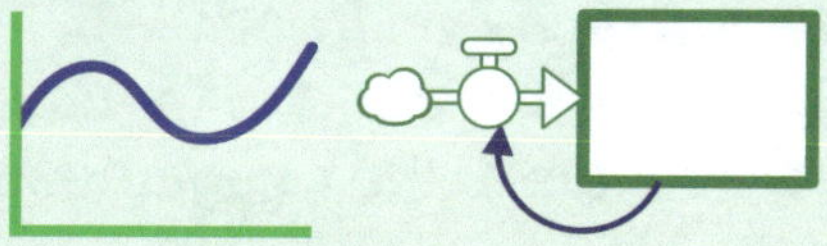

Skill

"People routinely fill and drain their bathtubs and manage their bank balances. Yet research shows that people have only weak intuitive understanding of stocks and flows." John D. Sterman

Overview

Recognizing how stocks and flows work is crucial to understanding the dynamics of a system. Stocks are accumulations that result from behavior over time. Flows are the rates that determine how those accumulations increase or decrease. Barry Richmond used the analogy that stocks are like nouns and flows are like verbs.

Because stocks can have both inflows and outflows, a policy or behavior that changes the rate of one flow may not have the desired or expected effect on a stock. For example, if you are pumping water out of a flooded basement (the outflow), the water level (the stock) won't go down until the rate of the water coming in (the inflow) is less than what you are pumping out. This example seems simple, but the principle applies to complex systems too.

After students learn about or review stock/flow thinking, they will tackle the "Department Store Task," originally created by John Sterman. Then they draw stock/flow diagrams about atmospheric CO_2 levels.

Materials

- Materials to introduce stocks and flows, p 203
- The Department Store Task instructions, p 204
- Diagramming atmospheric CO_2 as a stock/flow diagram, p 204
- Handout: The Department Store Task, p 110
- Resource: Intro to BOTGs, p 215
- Resource: Intro to Stock/Flow, p 223

Looking Back

Lesson 1.4 Recycling introduced students to the bathtub analogy for stocks and flows. If students did not complete Lesson 1.4, use steps #1 – 5 in that lesson to introduce the basic concepts of stock/flow thinking.

Procedure

1. Review what students know about stocks and flows.
2. Students with less experience using stock/flow diagramming will benefit from practice using the resources listed under Materials.

3. Distribute "The Department Store Task" handout at the end of this activity plan. You can read about it in the blog post link under Materials. Be careful not to reveal the solutions!

   First, have students answer the questions independently. Then have them discuss their answers in teams and see if they reach consensus on their choices and can explain their reasoning.
4. Open the conversation up so that the teams see if they agree on their answers. Challenge them to find a general rule that can be used to determine the quantity of a stock given the data from flows.
5. Reveal the answers and ask students to explain the reasons for the correct responses. See Bringing the Lesson Home for guiding questions.

"Trust the kids"

Teachers who use systems thinking tools and ideas soon discover that students generally pick up the concepts quickly. Veteran systems educators advise those new to the field to "Trust the kids." Young people will usually master the software quickly, draw connection circles and causal loop diagrams confidently (albeit with missteps at first), and work collaboratively. School is a social hub for most kids, so teamwork appeals to their eagerness to interact.

Many students are veteran video gamers. They won't be intimidated by software and can bounce back from mistakes quickly. Most systems educators find they can move instruction much more quickly with student groups than during adult workshops. It makes sense to "Trust the kids."

Create a stock/flow diagram for atmospheric CO_2 (FIGURE 1). Ask students to draw the stock and then add flows.

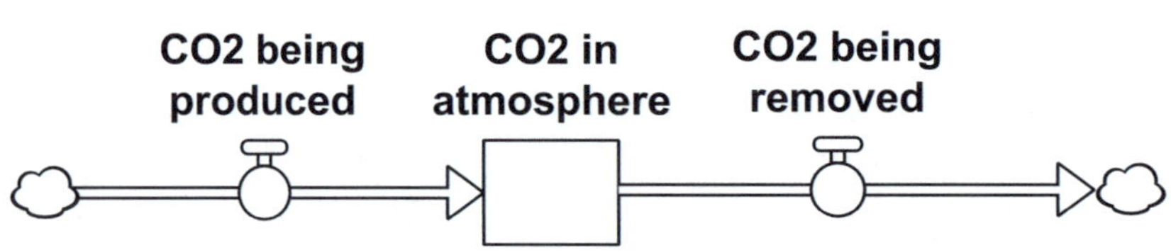

FIGURE 1

Let them discuss in teams what the stock/flow diagram explains about the amount of CO_2 in the atmosphere. If the inflow of CO_2 is greater than the outflow, the amount in the atmosphere will increase. Even if people can manage to reduce the production of CO_2 caused by fossil fuel combustion (FIGURE 2), the level in the atmosphere will continue to rise until the outflow exceeds the inflow.

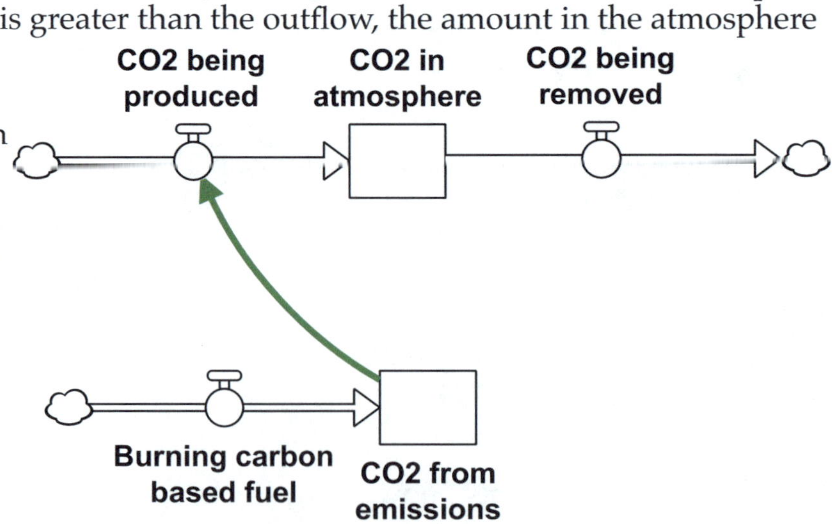

FIGURE 2

And as long as the atmospheric CO_2 remains very high, the radiation of Earth's heat into space will be less than the heat received from the sun.

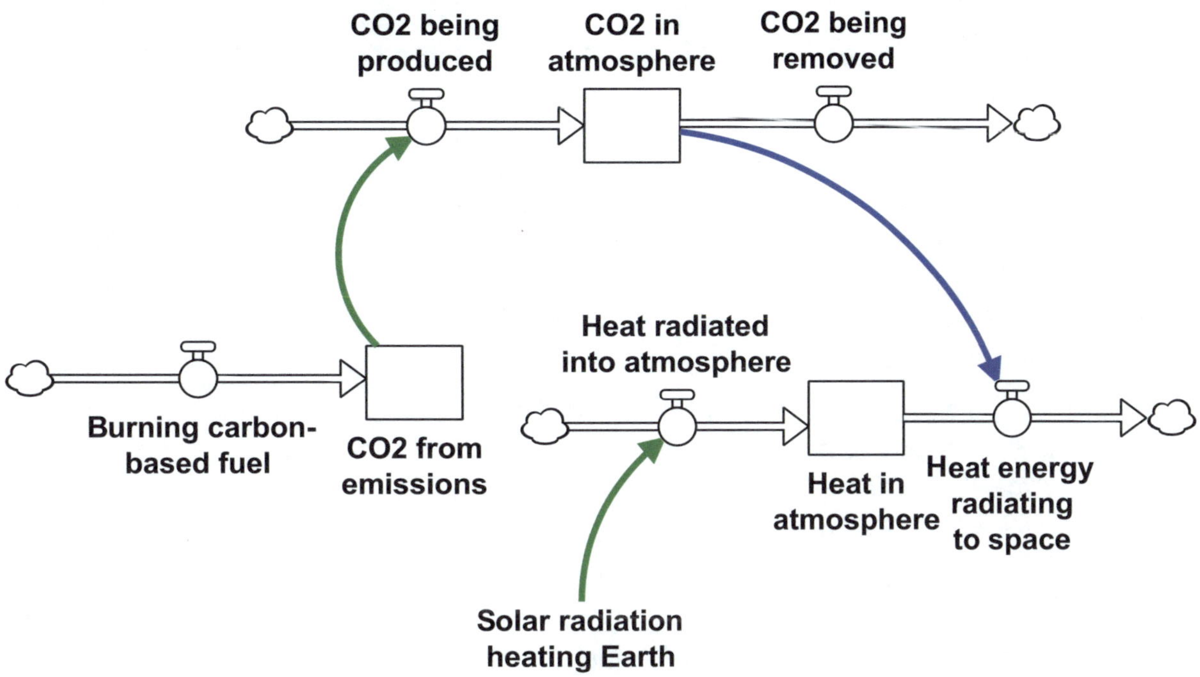

FIGURE 3

Bringing the Lesson Home

Why is a problem like the department store task tricky for many people?

The number, or stock, of people in the store when the graph starts is not able to be determined. The graph is a record of inflow and outflow into and out of the store. It does not record the number of people in the store but rather the change in that number. The flows allow us to tell when that stock of people in the store increases or decreases, and at which times during the 30-minute period the number is greatest and smallest. To find the times when the most and fewest people were in the store, you need to consider the difference between the flow of people entering and the flow of people leaving.

How were you able to solve the problem?

Questions 1 and 2 – These tend to be the easiest for people to answer. Finding the highest point on the line of people entering the store tells when the most people entered during one minute. Finding the highest point on the line of people leaving the store each minute told when that flow reached its maximum.

Many people find Questions 3 and 4 difficult. Have students who have thought out the solution explain it to others in their own words.

Bringing the Lesson Home
(continued)

Question 3 –When the inflow of people entering was greater than the outflow of people leaving, the total people in the store increased. That happened during every minute through minute 13, when the inflow line was higher on the graph than the outflow line. When the graph lines for the inflow and outflow crossed, more people were leaving than entering, and the number of people started to decline. Therefore, that's when the most people were in the store.

Question 4 – After minute 13, more people left the store every minute than were entering. To find out if the total number of people in the store was smaller at the end of 30 minutes compared to the beginning of the period, we needed to determine if more people left or entered the store during that time.

Some students may have calculated the net change during each minute and concluded that more people left than entered during the 30 minutes. Others may have used a more visual approach with the graph to avoid calculations and answered more quickly. The area between the two lines represents the change in the people in the store. When the inflow line is higher than the outflow line, the number of people in the store increases by the difference between them. When outflow is higher, the number of people in the store decreases by the difference.

The graph clearly shows two different sections. During the first 13 minutes, the space between the lines indicates that the net change was positive – more people each minute. After 13 minutes, the area between the lines indicated a negative net change – fewer people in the store after each minute. Because the area in the later section of the graph when more people were leaving is larger than the area in the first section, we can tell that overall more people exited the store during the 30 minutes than entered. Therefore, the fewest people were in the store at the end of the 30-minute period.

Can you think of situations when the ability to understand the behavior of a stock by knowing the flows would be important?

Stocks only change because of flows. The change in quantity of any stock that has both inflows and outflows is determined according to the net amount of those flows. Students will be able to cite examples on a variety of levels.

- *The money in a piggy bank (stock) is added to by deposits (inflow) and reduced by withdrawals (outflow). If a person puts more money in than is taken out, the total grows.*
- *Credit card debt (stock) increases when we charge purchases (inflow), as well as when the company charges interest (another inflow). Debt can be reduced by paying back (outflow).*
- *A person's confidence level can increase as one gains confidence (inflow) and can decrease when one loses confidence (outflow).*

Bringing the Lesson Home
(continued)

If the net differences among flows is not understood or calculated properly, the behavior of the stock may not satisfy our goals. Looking only at some of the flows in a system can provide people with an incomplete picture. For example, a government may reduce the size of the yearly deficit in its budget but still allow its debt to rise. A deficit means that money being spent (outflow) is greater than money received as revenue (inflow). If there is a deficit, the debt (stock) owed by the government will rise, even if the current deficit is smaller than the previous year's.

How does your simple stock/flow diagram help explain the problem of increasing amounts of atmospheric CO_2?

Reducing the amount of CO_2 entering the atmosphere as a result of fossil fuel combustion (inflow) is an important goal. But the amount of CO_2 present in the atmosphere (stock) will continue to rise until the removal (outflow) equals or exceeds the inflow.

If the inflow of CO_2 is greater than the outflow, the amount in the atmosphere will increase. Even if people can manage to reduce the emissions of CO_2 caused by fossil fuel combustion, the level in the atmosphere will continue to rise until the outflow exceeds the inflow.

Handout: The Department Store Task

Figure 1. "Classic" Department Store Task (Sterman, 2002; Cronin, Gonzalez, and Sterman, 2009).

The graph below shows the number of people *entering* and *leaving* a department store over a 30-minute period.

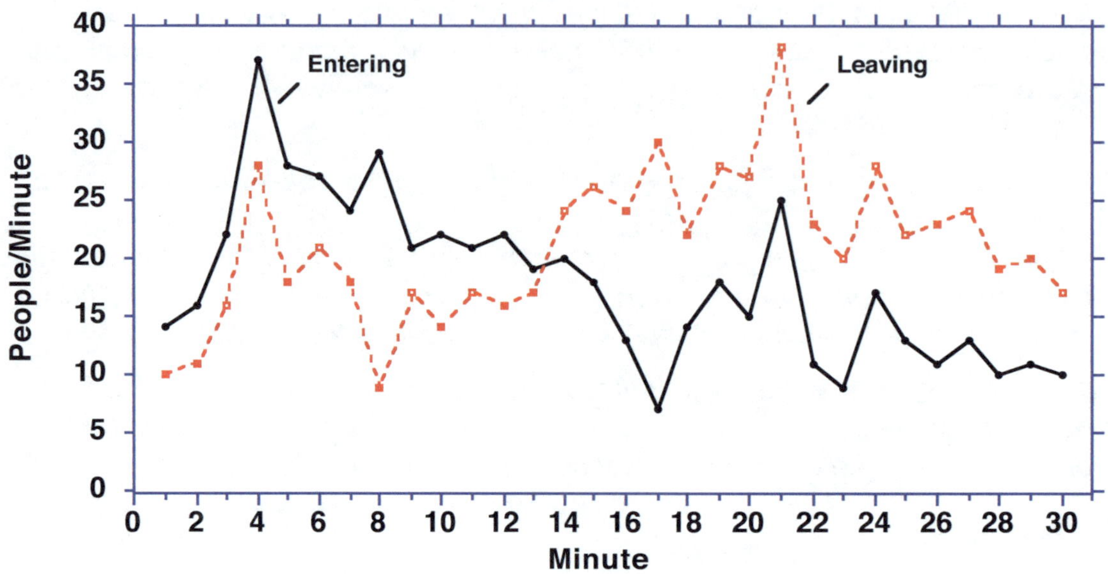

Please answer the following questions.

Check the box if the answer cannot be determined from the information provided.

1. During which minute did the most people enter the store?

Minute ________ ❑ Can't be determined

2. During which minute did the most people leave the store?

Minute ________ ❑ Can't be determined

3. During which minute were the most people in the store?

Minute ________ ❑ Can't be determined

4. During which minute were the fewest people in the store?

Minute ________ ❑ Can't be determined

3.4 The Carbon Cycle
Modeling Where the Carbon Goes

Environmental Systems
Why is carbon accumulating in the atmosphere and why does that matter?

Tools

Skill

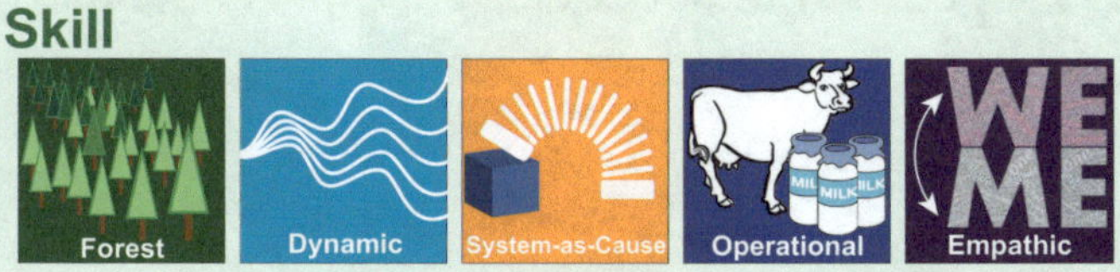

"When we try to pick out anything by itself, we find it hitched to everything else in the universe." John Muir

Overview

Carbon is one of the most common elements on Earth and constitutes a basic building block of life. It bonds with other elements to form many types of molecules, and it exists in an essentially closed system on our planet. The total amount of carbon remains practically constant, but measurement and scientific investigation have shown that carbon dioxide (CO_2) levels in the atmosphere have been rising since the late 1700s. Amounts are now higher than any time in at least the past million years or more.

In this activity, students will create a stock/flow diagram to model the carbon cycle, tracing the movement of carbon in Earth's systems. After they investigate the effect of atmospheric CO_2 on global climate, students will present their findings. Understanding the carbon cycle is crucial to formulating climate change policies.

Materials

- Carbon cycle and greenhouse gas information, p 204
- Instructions for building diagrams and models, p 204
- Diagramming and modeling materials
    - ◊ For diagramming, no materials necessary; diagrams may be drawn by hand.
    - ◊ For computer drawing and model building, computers with system dynamics software
- Resource: Intro to BOTGs, p 215
- Resource: Intro to Stock/Flow Diagrams, p 223
- Resource: Intro to System Dynamics Computer Models, p 224
- Resource: Intro to System Dynamics Modeling Software, p 238

Procedure

1. Introduce students to the Keeling Curve, either through their own research or by projecting an image in class. Be sure they understand that this behavior over time graph shows the concentration of CO_2 in the atmosphere over time from 1958 to the present. Look at graphs that

show atmospheric CO_2 over longer time scales also.

2. Where is the additional carbon in the atmosphere coming from? Have students draw conceptual diagrams of the carbon cycle. Examples available online or in texts vary in complexity. Students' prior knowledge of the carbon cycle will determine which diagrams are most accessible for them.

   How can the carbon cycle be represented through a stock/flow diagram? Challenge student teams to create stock/flow (S/F) diagrams to show how carbon enters and leaves the atmosphere. If students have never used S/F diagrams, introduce them by following steps #1-5 in Lesson 1.4. Even for experienced students, it can be beneficial when the whole class or small teams can collaborate to build models together.

3. Students who are building working models can find equations on the website linked in the Materials section. Students who are drawing S/F diagrams rather than models can use the following sequence. This diagram is simpler and different from the working models linked under Materials. All students can run the En-ROADS simulations using the Climate Interactive website.

Long-term Feedback Loop

In her book, Timefulness, Marcia Bjornerud describes how quantities of carbon sank to the ocean floor during periods of extensive glaciation in Earth's history. Windy conditions blew nutritious "dust" into the oceans, allowing phytoplankton to flourish. After removing atmospheric CO_2 through photosynthesis, some phytoplankton sank to the bottom before decomposing. Less atmospheric CO_2 continued to cool global climate, and glaciers expanded their range. Students may represent this process with causal loop diagrams, (See FIGURE 1).

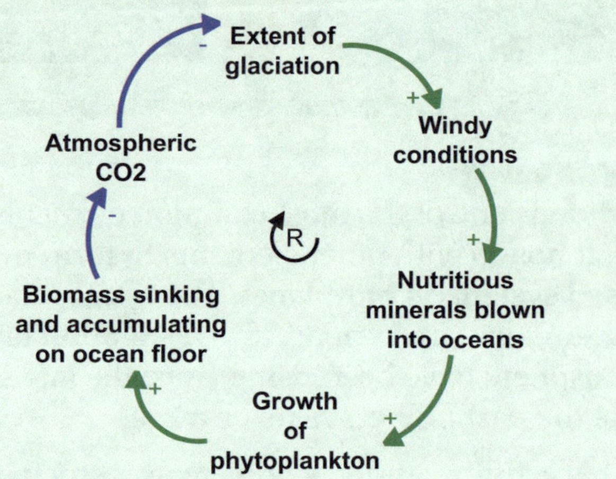

FIGURE 1

What are the stocks? Where is carbon stored in, on, and around Earth (FIGURE 2)?

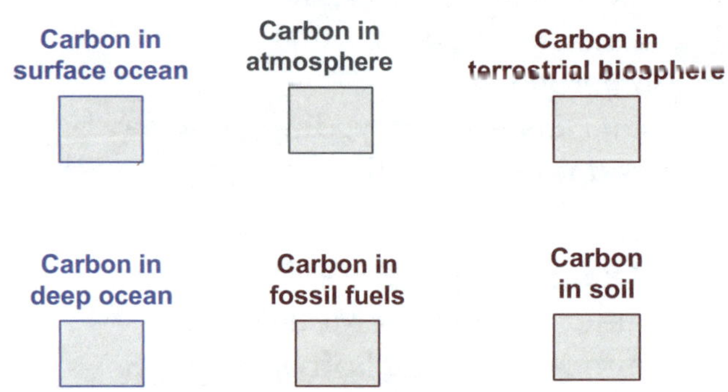

FIGURE 2

Students should identify at least six places where carbon resides on Earth. "Fossil fuels" can include other non-living material in Earth's lithosphere. More carbon is stored in deeper layers of Earth, but in this simplified diagram, we have omitted those stocks that move much more slowly. However, students will find that human activity affects even long-term processes. For example, when organisms die and sink to the ocean bottom, carbon can build up over time. As ocean chemistry changes, ocean biomass may be reduced, allowing more carbon dioxide to remain in the water and air.

What are the flows in the carbon cycle? By what processes does carbon move among the stocks (FIGURE 3)? Add the flows, one by one. Some system dynamics software programs include a "storytelling" feature that allows a modeler to reveal diagram elements sequentially when sharing.

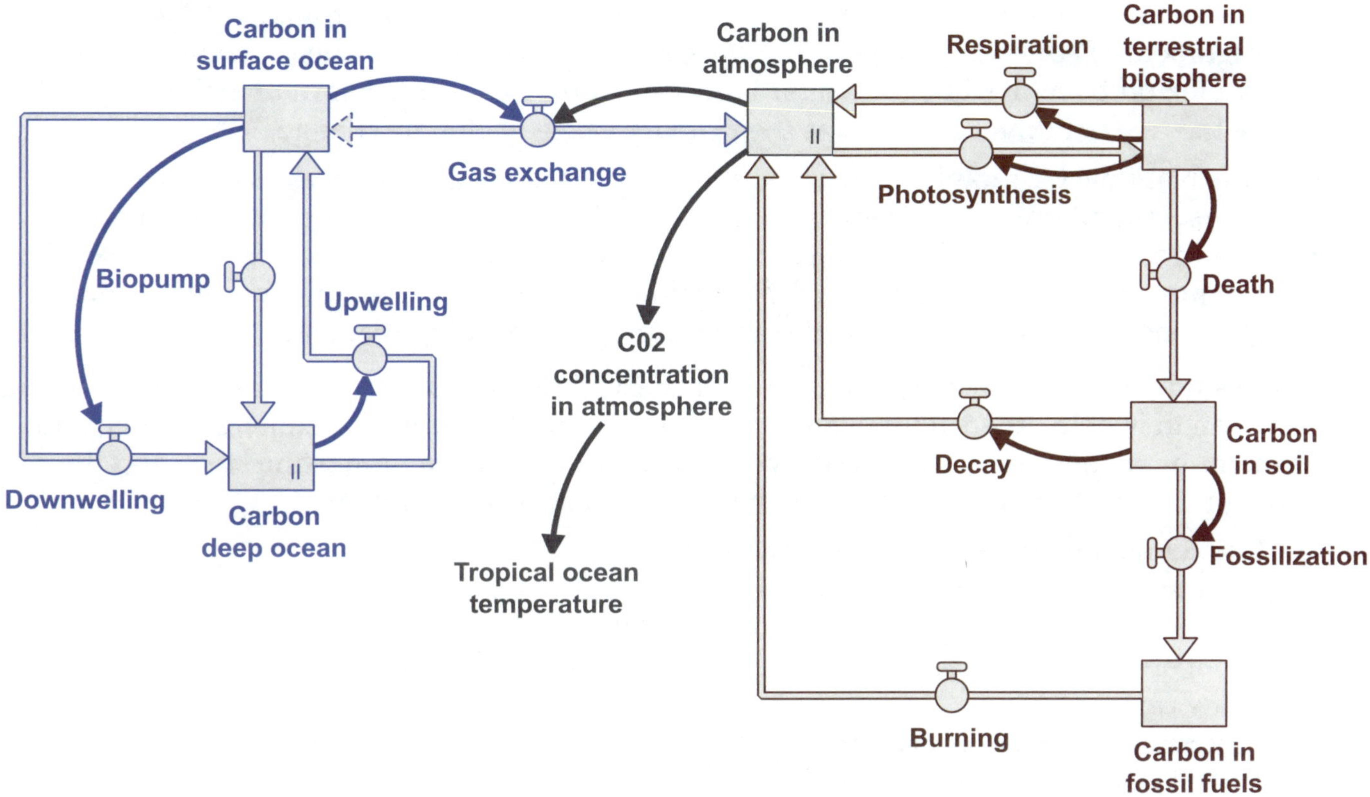

FIGURE 3

Note that the flow "Gas exchange" between "Carbon in surface ocean" and "Carbon in atmosphere" is a biflow. It has arrows on each end indicating that carbon can move in either direction. Carbon may move between the other stocks, but those are drawn/labeled as separate inflows and outflows to provide more clarity.

4. After students complete their stock/flow diagrams of the carbon cycle, have them analyze why CO_2 is increasing in the atmosphere. They should see that fossil fuel burning adds carbon to the atmosphere. Human activity since the Industrial Revolution has drastically increased carbon emissions. Photosynthesis removes CO_2 from the

CO_2 levels in the atmosphere

The amount of CO_2 in the atmosphere has varied over the history of Earth. But the concentrations in the air in the 21st Century are believed to be higher than at any time in at least the past 800,000 years. Modern levels may match those of 3 million years ago when Earth's temperature and sea level were significantly higher. CO_2 is a major factor in trapping heat and therefore raising global temperatures. Reducing the concentration of CO_2 in the atmosphere will be necessary to combat climate change.

atmosphere as plants grow and store carbon. Deforestation and other processes that remove plants from the biosphere reduce the removal of atmospheric CO_2 through photosynthesis.

5. As teams further analyze their diagrams and models, ask them to propose ways to address the problem of increased CO_2 in the atmosphere. How can they educate a variety of audiences? Have students form consulting groups. Their job is to educate people about the carbon cycle – what it is, how humans are affecting it, and how carbon is entering the atmosphere faster than it is being removed.

   Students can use their understanding of the carbon cycle and systems thinking tools to tackle the assignment of educating the public. How can stock/flow thinking help people understand the complex problem of climate change? How can students address alternative theories that dispute the role of human activities as the primary cause of climate change?

6. Show the National Geographic materials, either projected for the whole class or via team computers. John Sterman, a system dynamicist at MIT, believes that many people do not understand stock/flow thinking. He is the creator of the Department Store Task in Lesson 3.3, which demonstrates how difficult stock/flow thinking can be. It's not that we aren't smart enough – most of us don't have practice and training for thinking that way.

   But the students have gained experience with stock/flow thinking and other systems thinking skills. Can they help others understand how the CO_2 problem works? Challenge students to use their understanding of the carbon cycle to present the information using stock and flow language. Suggest some options for them to use, and brainstorm others:

   ◊ Deliver an illustrated lecture presentation.
   Have students imagine that a middle school science teacher asked them to explain the carbon cycle to her class. Or, could they explain it to an elementary school class? How about an audience of adults who are not professional scientists?
   ◊ Draw cartoons.
   Editorial cartoons are very effective and will appeal to many students who enjoy drawing and exercising their sense of humor. Because the prior knowledge among your students will vary, you need to make a teaching decision about presenting examples before they create their cartoons.
   ◊ Prepare a slide show.
   Presentations may include cartoons and graphics.
   ◊ Create and perform a song, rap, or poem.
   ◊ Write an explanation in their own words, integrating the stock/flow model.
   Students can write persuasive essays, structuring them as "op-ed" pieces, or letters to the editor of newspapers and periodicals.
   ◊ Prepare and deliver a presentation to a town meeting.
   Imagine your community needs to decide whether or not to use the resources of the local government to create and promote programs aimed at limiting carbon emissions. Advocate for policies government should implement. Some cities and towns are moving toward renewable energy sources to power schools, municipal offices, and other public facilities. Explain why this is important.

7. After student teams practice and present to each other, find potential audiences for their presentations, or videotape them for possible distribution. If you can locate audiences for student presentations, the experience will be more meaningful.

Bringing the Lesson Home

What causes an increase in CO_2 in the atmosphere?

Carbon is always moving in the Earth's carbon cycle. But more stored carbon is being released than ever before, changing from terrestrial storage underground to gas in the atmosphere. When coal, petroleum, and natural gas are burned as fuel, the carbon combines with oxygen and enters the atmosphere.

Plants remove CO_2 from the atmosphere through the process of photosynthesis. Over time, humans have cleared plants and caused the rate of removal to slow. In short, more carbon is entering the atmosphere than is being removed.

Why is the level of carbon in the atmosphere at such high levels compared to other periods in Earth's history?

Normally the carbon cycle balances the stocks of carbon, but since the Industrial Revolution, humans have burned large quantities of stored carbon. That combustion causes carbon to combine with oxygen to form CO_2. The CO_2 stays in the atmosphere until it is removed. Plants provide a major source of carbon dioxide removal as they perform photosynthesis and store carbon. But human population has grown at an increasing rate, causing forests to be cleared and in some cases burned, reducing the amount of photosynthesis occurring.

Are the models completely accurate? Why are there differences among versions of CO_2 models?

Statistician George Box wrote, "All models are wrong, but some are useful." The carbon cycle is an especially complex system, occurring on a gigantic scale and presenting numerous challenges for modelers. Laboratory experiments cannot reproduce the physics of an entire planetary system. The carbon cycle contains long delays, and separating out human-caused factors from natural cycles requires the inclusion of a range of probabilities.

How can we explain an increase in atmospheric carbon using a stock/flow diagram?

Carbon flows into the atmosphere from both human and non-human sources, and is removed primarily by plants. Since the Industrial Revolution, combustion of wood and of fossil fuels has increased the inflow while deforestation has reduced the removal outflow. When inflows are greater than outflows, stocks increase. In order to reduce the amount of carbon in the atmosphere, the outflow must be greater than the inflow.

Stocks can only be changed by flows. The simple one-stock diagram of CO_2 in the atmosphere shows that until the outflow is greater than the inflow, the amount of CO_2 will continue to increase. Even if we reduce the amount of CO_2 emissions that make up that inflow, the stock will not decrease until the inflow is lower than the outflow. In the first two decades of the 21st Century, emissions have continued to rise steadily with increases almost every year.

Bringing the Lesson Home
(continued)

Based on your stock/flow diagram or model, what policies would you suggest to stop the rise in atmospheric CO_2?

Because the outflow must exceed the inflow to reduce the stock of atmospheric carbon, students will suggest ways to change both flows.

Two key flows in atmospheric carbon are the burning of fossil fuels (inflow) and the removal of carbon through photosynthesis (outflow). The process of fossilization occurs on a very long timescale while burning is immediate, so keeping fossil fuels in the ground will slow the inflow of atmospheric carbon. Plants provide the primary outflow for removing atmospheric carbon. Therefore, preserving and expanding forests and other plant habitats could reduce the stock of atmospheric CO_2.

The stock/flow diagram of the carbon cycle allows us to see where increasing carbon is coming from and where it might be removed. The burning of fossil fuels produces most of the imbalance by increasing the inflow of CO_2, combined with the continuing removal of forests that is decreasing the outflow.

Some of the policies necessary to reverse the increase in atmospheric CO_2 may be costly and require fundamental changes. How can people learn about what needs to change and what they can do? What systems thinking tools and ideas might be helpful?

Complex problems are difficult to understand and never have simple solutions. Reducing our reliance on fossil fuels can impact many people's everyday lives. We must commit to short-term changes in order to insure long-term sustainability. Systems thinking diagrams can help convince people that current practices are unsustainable and change is necessary.

BOTGs of CO_2 in the atmosphere matched to rising average temperatures clearly show the connection between fossil fuel use and climate change. Stock/flow diagrams of atmospheric CO_2 present convincing evidence that small reductions in emissions alone will not reverse those trends.

3.5 The Greenhouse Effect
Trapping the Heat

Environmental Systems
What is the greenhouse effect?

Tools

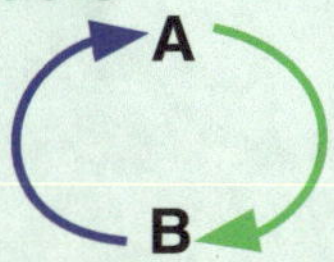

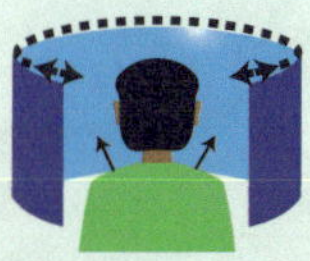

Skill

"Fire made us human, fossil fuels made us modern, but now we need a new fire that makes us safe, secure, healthy and durable." Amory Lovins

Overview

Some students will have the impression that 'the greenhouse effect' is a bad thing. But without the Earth's atmosphere trapping some of the heat radiated from the surface of the planet, we wouldn't have the relatively warm environment that has allowed life to flourish in so many diverse forms.

Smaller scale examples of the greenhouse effect can also demonstrate its positive aspect. Greenhouses allow people to grow and protect organisms that otherwise would not survive at lower temperatures. A closed car parked in the sun may be dangerously hot in summer, but can feel like a welcome sanctuary when outside temperatures fall toward freezing and below. Houses and other buildings can often be oriented on their sites in ways that allow for passive solar heating during colder months.

What IS a bad thing is when the greenhouse effect traps so much heat that temperatures on Earth become too warm. While CO_2 receives most of the attention, methane and other gases can also contribute to overheating the Earth.

Students build and experiment with mini-environments to model the greenhouse effect and relate their outcomes to Earth's climate system.

Materials

- Building materials for greenhouse containers
  - ◊ Thermometers
  - ◊ Containers like shoeboxes, flower pots, storage containers, trays, etc.
  - ◊ Plastic wrap
  - ◊ Variety of materials to simulate surfaces on Earth - water, sand or dirt (e.g., potting soil), rocks in a variety of colors, flour, and other materials students choose
- Media about the greenhouse effect, p 204
- Handout: Greenhouse Effect, p 120
- Resource: Introduction to CLDs, p 220
- Resource: Introduction to Systems Simulations, p 225

Procedure

1. Watch videos or read articles describing the greenhouse effect to gain a further understanding of climate challenges. The NPR audio story about methane emissions ends with a description of a simple reinforcing feedback loop that students can draw and discuss (FIGURE 1).

2. Distribute the handout and explain that each team will use containers and a variety of materials to build small environments that test the greenhouse effect. Working together, students will demonstrate how solar energy is trapped and the effect of different surfaces. Depending on the class's level of scientific experience and background, you may have to review the procedure for designing a valid experiment.

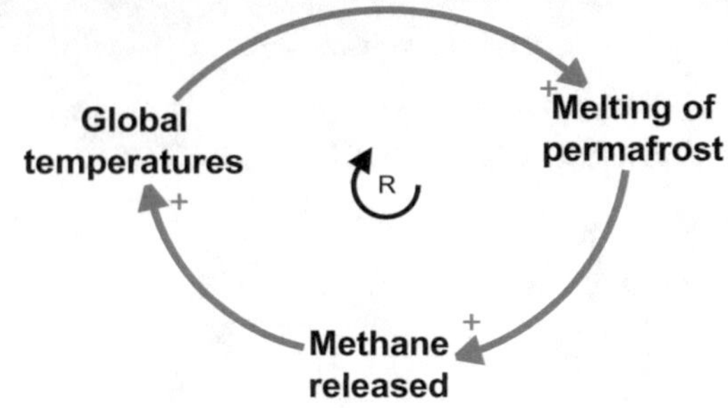

FIGURE 1

3. Encourage a broad variety of experiments. Every team should receive a pair of similar containers that they will fill with the material they want to test. Cover one container with transparent plastic wrap and leave the other open. Use lamps with incandescent bulbs to heat up the containers carefully, or place them in the sun, either outside or on a windowsill. Compare the temperatures in the two containers.

4. Then students can cover both containers and compare different interior surfaces. They can choose materials that simulate land, bodies of water, reflective and absorbent colors, and a variety of soils. As a class, students can brainstorm a range of variables to test in order to simulate aspects of the greenhouse effect. Examples of simulations include -
   - atmospheric blanket effect – compare temperatures in containers when the plastic is perforated versus sealed
   - reflectivity – is there a difference in temperature among the containers when different colored materials are placed inside?
   - conductivity of materials – leave the bulbs of the thermometers in each material after removing the heat source and measure how much heat is stored in each case. Remove the cover and measure the rate of cooling for different materials. Compare cooling curves of various materials.
   - temperature differences in various conditions – example: humid versus dry environment, spraying water in one closed container and leaving another dry

5. After the experiments, have each team present their conclusions as a slide show, poster presentation, or other report.

Designing an experiment

Most students will be familiar with some form of the "scientific method" but a review never hurts. Use a reputable website for review of the scientific method before starting. The specific sequence of steps in an experiment will vary depending on what source you cite but be sure students know how to set up a reliable experiment. Urge them to pose a question or hypothesis and then carefully control variables to get a valid result.

Bringing the Lesson Home

How would you describe the greenhouse effect in your own words?

Energy from the sun penetrates the atmosphere and warms the surface of the Earth. Heat is given off as infrared radiation. The atmosphere acts like a blanket over the Earth and keeps some heat from escaping into space. The trapped heat keeps the temperature of the Earth higher than it would be without an atmosphere.

What does your experiment simulate about the greenhouse effect?

The container represents the Earth and the plastic over the top simulates the atmosphere. The material in the container represents the Earth's surface. Depending on individual experiments, students will have tested different variables.

Is the greenhouse effect a good or bad thing?

The greenhouse effect is necessary for our survival. Without our atmosphere to modulate temperatures on Earth, our planet would be extremely barren. The diversity of life on our planet would be drastically reduced to very few life forms, if any. We need the blanketing effect of the atmosphere to keep us warm, and to keep temperature fluctuation within a range manageable for many organisms.

The bad side of the greenhouse effect occurs when the gases in the atmosphere trap too much heat and threaten to raise the temperatures on Earth to levels that endanger life on the planet. Conditions can change, making it difficult or impossible for species to survive. Severe weather events and climate change might cause catastrophic damage over time.

Handout - Greenhouse Effect

Type of Experiment: (circle one)	atmospheric blanket reflectivity conductivity temperature Other:____________________________
My Plan:	
Observations with dates/times:	
My Conclusions:	

3.6 Global Warming
The Human Effect

Environmental Systems
What effects are we feeling from rising global temperatures?

Tools

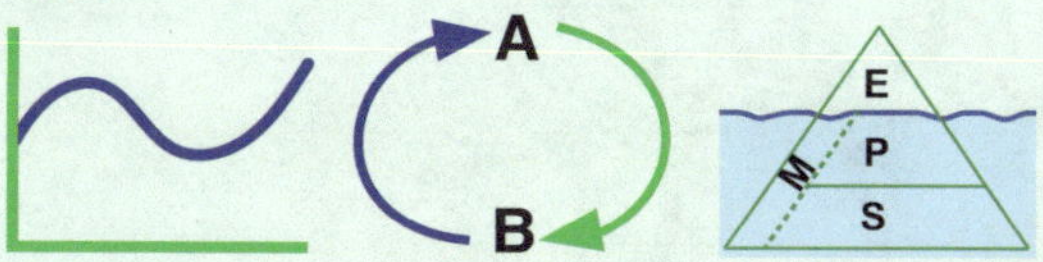

Skill

"Perhaps the time has come to cease calling it the 'environmentalist' view, as though it were a lobbying effort outside the mainstream of human activity, and to start calling it the real-world view."
Edward O. Wilson

Overview

Average global temperatures continue to increase significantly, clearly shown by behavior over time graphs that slope upward steadily. Scientists are analyzing the reasons for the warming and the consequences that may develop. Two vital questions are:

1. How much responsibility do humans bear for the warming temperatures the Earth has experienced?
2. What are the consequences of global warming?

Students investigate evidence of the effects of global warming. They summarize their findings using iceberg diagrams and feedback loops.

Materials

- Historical behavior over time graphs, displaying temperature and CO_2 levels, p 205
- Introductory videos and articles about global warming, p 205
- Handout: Iceberg Template (optional), p 227
- Resource: Intro to BOTGs, p 215
- Resource: Intro to CLDs, p 220
- Resource: Intro to the Iceberg, p 226

Procedure

1. Choose and share behavior over time graphs of global temperature over time (FIGURE 1). Ask four questions about the graphs to have teams analyze and discuss the data:
   - What is changing? Identify the variable(s) being tracked.
   - How is it changing? Describe the rate and direction of change.
   - Why is it changing? What are the causes?
   - So what, now what? What is significant and how does that affect policy?

Tell students they will research aspects of global warming and apply the same four questions to their specific topics.

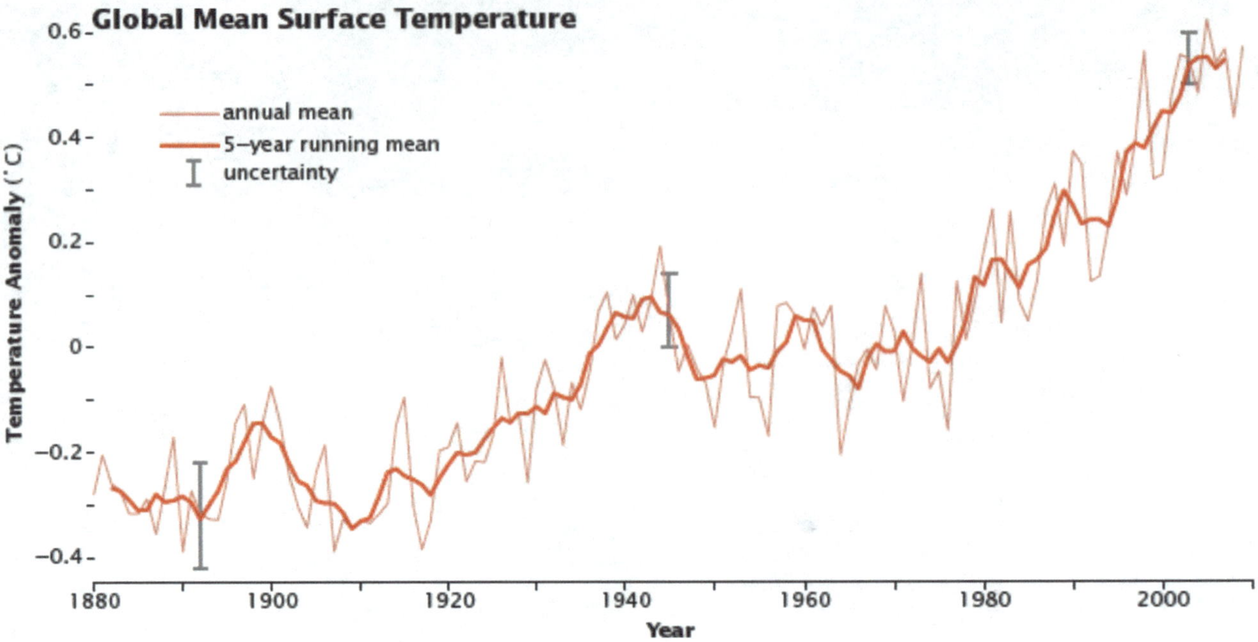

FIGURE 1

2. Show an introductory video about global warming. Organize teams and assign articles, or have them find resources independently.

   During their research, students may find arguments against human activity as the major cause of climate change. These positions should be examined along with the majority of prevailing research, but warn students to check the legitimacy and credentials of the individuals and the sponsoring organizations.

   It is important to find data that are supported by research and accepted by significant numbers of scientists. Students should try to discover what areas of consensus exist and why.

Global Warming Skeptics

By 2015, a large majority of scientists strongly believed that human activity was contributing to climate change in a major way.

http://climate.nasa.gov/scientific-consensus/

Contrarian theories continue to be defended, as is the case in much scientific work. Remind students that scientists do not deal with absolute certainty but rather propose explanations that fit observations and design experiments to test hypotheses. Experimental results require repeated testing, but world climate cannot be reduced to controlled variables in a laboratory. Facts are not final, and what is accepted as true today may be disproved in the future. Many websites provide arguments countering those who deny human activity is changing climate.

3. Have each team prepare a short summary or extract of their article. Require an iceberg diagram and/or a causal loop diagram (CLD). See the optional iceberg template. Examples of icebergs and CLDs are provided below for some of the suggested articles for guidance. Be sure to encourage students to create their own diagrams – working in teams should promote dialogue and careful analysis.

   Direct students to consider the present-day effects of rising global temperatures and address the fourth pair of questions in Step 1 – What is significant and how does that affect policy?

How do the current conditions they researched affect policy decisions for both short-term and long-term sustainability?

4. Decide on a format for sharing student research and systems thinking diagrams. Ideas for poster presentations and reports include:
   - Your team works for a political candidate as science consultants. Prepare a presentation that your candidate can use on the campaign to convince voters to support sustainable policies.
   - The administration of your school is considering investing in energy saving improvements. The school board seems skeptical about allocating funds. How can you as students help them decide to spend the money?
   - Some elementary classes in your town are studying the effects of global warming. Their teachers have asked if you could present an explanation of how the effects of rising temperatures are affecting the Earth (and maybe teach some beginners about systems thinking). Can you use feedback loops to explain global warming to younger students?
   - Research the arguments of people who contend that human activity is NOT the cause of rising temperatures on Earth. What scientific data are they presenting? What are the rebuttals to the case for rejecting human activity as a cause of rising global temperatures? What do you think?

Notes on the suggested articles with examples of iceberg and CLDs

In the hand-drawn generic iceberg diagram below, the mental model section considers global warming a primary cause of extreme weather observed in the events level. The structures level has a simple causal loop diagram describing feedback that drives more heat into the atmosphere. The patterns level uses behavior over time graphs to display changes on a longer time scale.

Most of the suggested articles in this activity will fit this iceberg format (FIGURE 2).

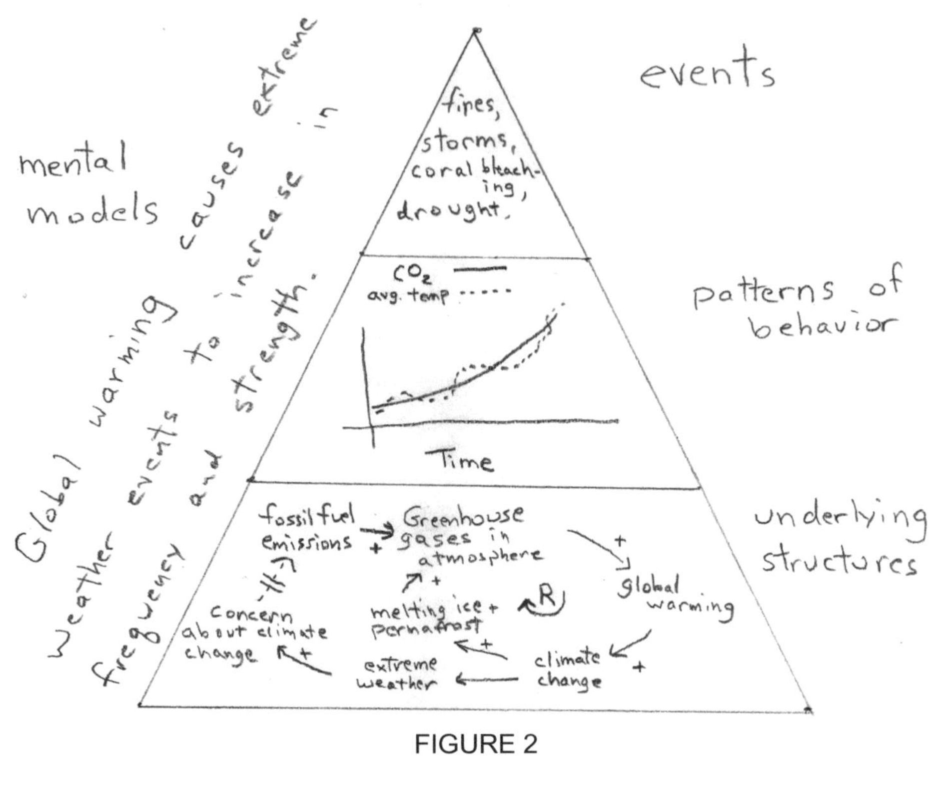

FIGURE 2

Article: "It's Not Your Imagination. Summers Are Getting Hotter"
This article uses behavior over time graphs effectively.

Article: "2018 is Shaping Up to be the Fourth-Hottest Year. Yet We're Still Not Prepared for Global Warming."
What are some effects of the rising temperatures around the world and what can be done about them?

Article: "A Historic Heat Wave Roasts Siberia"
In June 2020, record heat and dry conditions caused problems in a region known for its cold climate. The Arctic is experiencing the effects of global warming more quickly and severely than other parts of the world.

Article (Opinion): "How We Know It Was Climate Change" (FIGURE 3)
Because this article is an "op-ed" and therefore an opinion piece, it's a good candidate for an iceberg diagram. Here's a sketch of an iceberg diagram with placeholders for the patterns and structures levels:

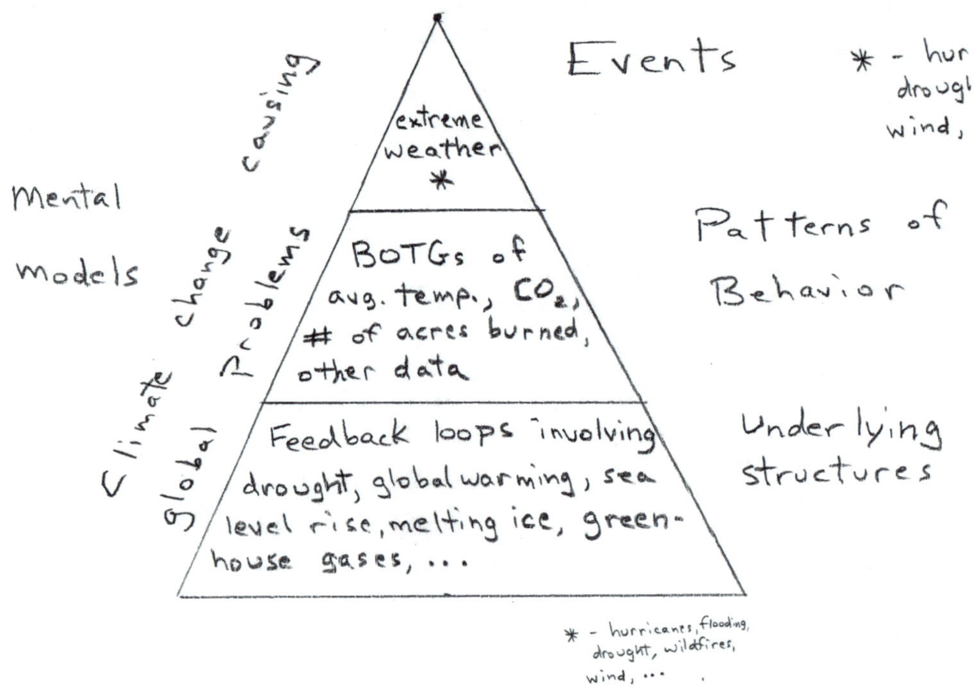

FIGURE 3

Article: "Arctic Sea Ice Missed a Record Low This Winter. Barely." (FIGURES 4 and 5)
The extent of ice in the Arctic Ocean shrinks each year

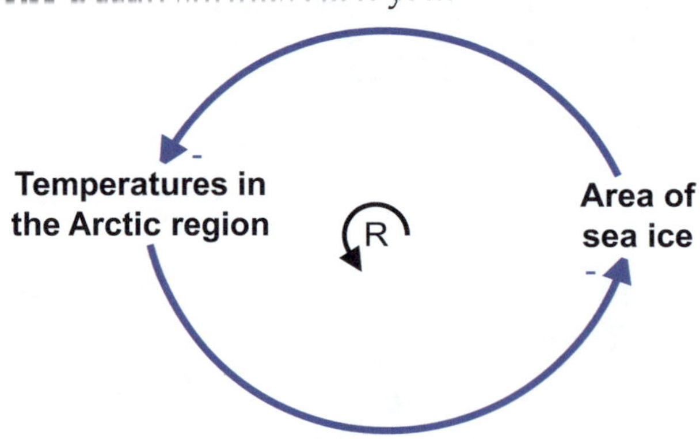

FIGURE 4

Notice how FIGURE 4 shows a reinforcing feedback loop even though both causal connectors are negative. Remember, in causal loop diagrams, a positive sign means "changes in the same direction," and a negative sign means "changes in the opposite direction." When the area of arctic sea ice goes down, temperatures in the region go up. When temperatures go up, the amount of ice goes down. This process results from the albedo effect, when areas of light color reflect more solar radiation than areas of dark color. So two negative or opposite causal connector arrows create a reinforcing feedback loop. Here's an expansion of that loop:

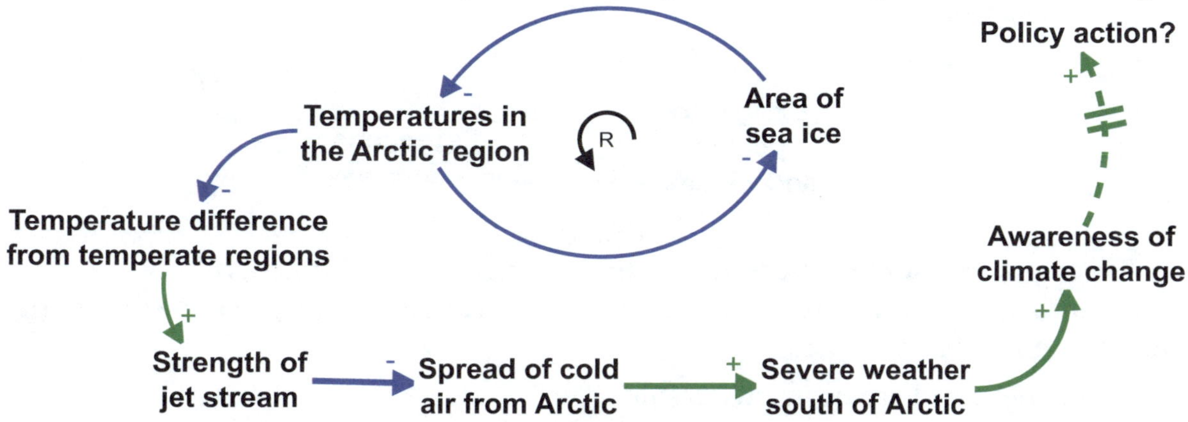

FIGURE 5

Will more severe weather cause enough concern to urge policy changes to curtail global warming? The two parallel lines indicate a delay and the dotted connector represents an unknown future course.

Article: "As Earth Warms the Diseases That May Lie Within Permafrost Become a Bigger Worry" (FIGURES 6 and 7)
In this article, scientists warn of potential dangers as warming temperatures in the far north allow disease to spread. As permafrost (ground that remains permanently frozen) melts, a reinforcing loop causes the release of CO_2 and methane, gases that trap heat in the atmosphere.

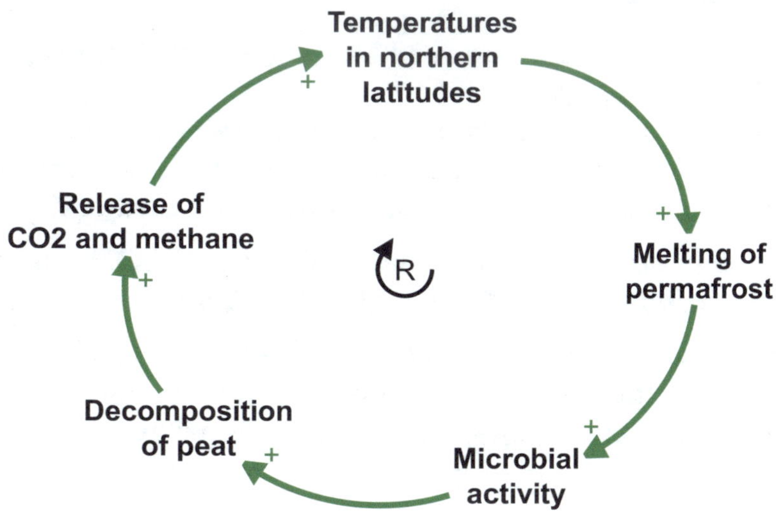

FIGURE 6

Dangerous pathogens also escape when permafrost melts. The causal loop diagram in FIGURE 7 uses a dotted connector to speculate on how the loop might be closed and reduce the threats to people.

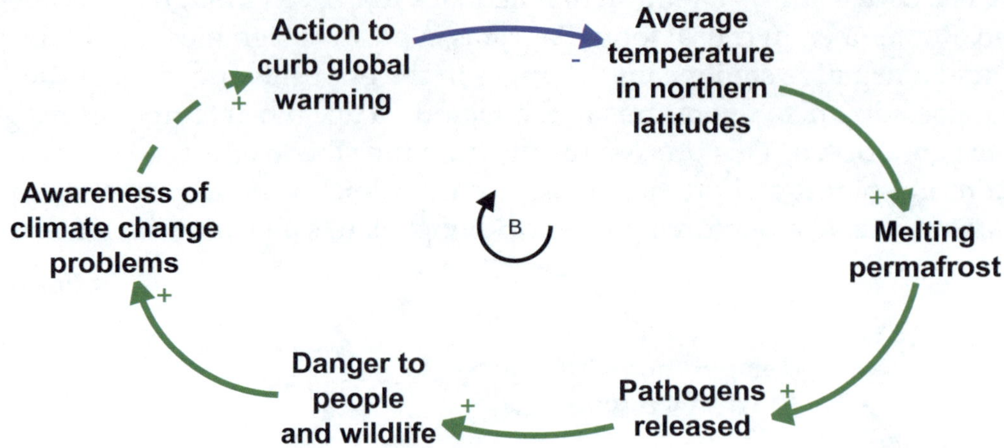

FIGURE 7

Article: "Portugal Forest Fires Worsen, Fed by Poor Choices and Inaction"
Multiple trends have contributed to deadly fires in Portugal. Among factors combining to produce strong feedback loops:

- Mismanagement and absentee ownership
- Fuel build-up
- Monoculture planting and increase in highly flammable trees
- Volunteer firefighters replacing trained professionals

Article: "How Global Warming Fueled Extreme Weather Events"
Five specific weather events in 2016 scientists believe were related to global warming:

- Record high temperatures
- Coral bleaching
- Drought in Africa
- Wildfires in North America
- Warm "blob" in Pacific Ocean

Articles: "In a Warming California, a Future of More Fire" and "Climate Change is Fueling Wildfires Nationwide, New Report Warns"
Dry, hot summers in California create conditions for severe wildfires. Wet seasons grow vegetation that shrivels, providing even more fuel during the following summer. Urban "heat islands" add heat to the atmosphere and reduce cloud cover, further drying out the vegetation.

Fires are more common in other US states, also. Wildfires add CO_2 to the atmosphere, making some forested areas contributors to additional global warming.

Article: "In Peru's Deserts, Melting Glaciers Are a Godsend (Until They're Gone)"
Shrinking glaciers threaten the water supply of people in Peru. Some of those who have migrated find opportunities reduced. Agricultural pests have expanded their range and reduced crop yields.

Article (Opinion): "What Climate Scientists Want You to See in the Floodwaters"
Describes why flooding events are more common and more severe, and what action can be taken to meet the challenges.

Bringing the Lesson Home

What are some of the arguments that claim human activity is causing the Earth's climate to become warmer?

Rising temperatures and CO_2 concentration in the atmosphere are highly correlated. CO_2 has been rising since the advent of the industrial revolution, when humans greatly increased the combustion of fossil fuels, releasing stored carbon into the atmosphere.

World population has increased exponentially. Forests have been cleared for land use, reducing the capacity of Earth's biosphere to remove CO_2 from the atmosphere.

What are other reasons temperatures might increase? What are scientists finding out about those possibilities?

Several theories suggest natural explanations for Earth's warming temperatures. Each theory has supporters and opponents. The Earth's climate system is extremely complex and many factors affect climate. Students should investigate non-human causal factors to better understand the arguments around climate change. These factors include:

- *Amount of energy received from the sun*
- *Tilt of the Earth*
- *Eccentricity of Earth's orbit*
- *Emissions from tectonic processes like volcanoes and ocean vents*
- *Randomness of data*
- *Time period under consideration – short-term versus long-term variations*

How can humans change their behavior to confront global warming?

Reversing the rise in global temperature will be a long and difficult process. Reducing the amount of greenhouse gases like CO_2 and methane in the atmosphere will only begin when more is being removed than is being added. Students will have suggestions based on their research and prior knowledge. Examples:

- *Transition from fossil fuels to renewable "green" sources.*
- *Protect and expand forests and undeveloped areas.*
- *Reduce consumption of disposable products.*
- *Increase efficiency of machinery, vehicles, lighting, appliances and other energy users.*
- *Construct new buildings and retrofit existing structures to save energy.*
- *Expand public transit and increase travel efficiency.*

3.7 Examining Climate Models
Working Together and Looking to the Future

Environmental Systems
How can the world's nations cooperate to address the climate crisis?

Tools

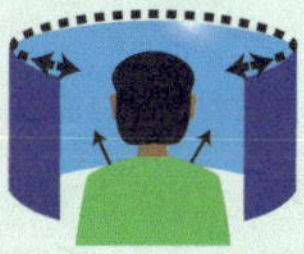

Skill

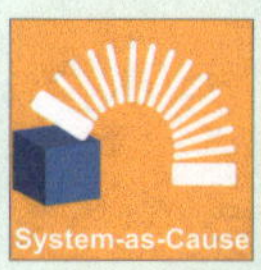

"Students should see modeling and an understanding of systems as a way to reduce social and political conflict."
Jay W. Forrester

Overview

Nations of the world must work together to meet the global climate emergency. Students assume the roles of negotiators and consultants as they participate in a system dynamics simulation based on a CO_2 climate model. They apply their knowledge of climate change, thinking about "What's changing, how, and why?" Using the simulation, they are encouraged to ask "So what?" and "Now what?" As students project climate models into the future, they learn what the consequences could be for a variety of different policies.

The simulated "climate summit" uses sophisticated computer models to test possible future scenarios. The Climate Interactive group has prepared a complete lesson plan for holding the summit. The steps in this lesson supplement their plan.

Materials

- Computers with internet access
- Climate background info, p 206
- Climate Interactive simulation website, p 206
  Be sure to register and familiarize yourself with the program ahead of time.
- Resource: Introduction to System Dynamics Computer Models, p 224
- Resource: Introduction to Systems Simulations, p 225

Procedure

1. Visit the Climate Interactive World Climate Simulation page. Familiarize yourself with the procedure and register to participate. Organize students into teams according to the Climate Interactive directions. They will simulate roles as representatives of different countries or as members of specific interest groups. Within that identity, they must decide what course of action is best to deal with climate change, both for their constituents and the world.

2. The Gapminder and EPA website links can orient students with an overview of the role the world's nations will have to play in meeting the climate crisis. Climate Interactive provides

excellent introductory material to prepare students for their role-playing in a climate summit. Their job will be to advocate for a policy agreement among nations that will address the problem of CO_2 levels in the atmosphere.

3. Teams will plan their advocacy strategies using the information from their earlier studies, prior knowledge, and data from the climate modeling websites. Opening presentations at the summit should explain their proposals and the rationale underlying their strategy.
4. During the negotiation phase of the summit, students attempt to forge agreements that will minimize global warming and avoid potentially catastrophic consequences from "business as usual."
5. Have each team make closing remarks and summaries as a debriefing exercise. Teams may respond to the positions of other teams after all presentations are completed. Inviting observers from outside your class to attend the summit can be valuable for generating questions and opinions from an audience.

Bringing the Lesson Home

In his work modeling the dangerous effects of CO_2 on climate change, Professor John Sterman has said, "Science is no longer the bottleneck. The challenge is political and social." What do you think he means?

Scientists overwhelming agree on the connection between atmospheric CO_2 and rising global temperatures. With such strong consensus, there is no question that policies must be implemented to address the problem. But the changes necessary require cooperation among nations and adjustments to people's lifestyles and patterns of consumption. Achieving changes on that scale is always difficult.

How are models useful in confronting the climate change dilemma?

Simulations can be run to predict what outcomes will occur given different scenarios. "If/then" kinds of questions can be posed for multiple situations, simulating the amount of CO_2 in the atmosphere and forecasting what the temperature, sea level, and other conditions will be at different points in the future. Predicting specific outcomes can influence public opinion and urge people to action.

How successful were the summit representatives in reaching an agreement? What factors were the most influential during negotiations?

It's important to have students reflect upon their experience. Reaching consensus on an action plan is complicated. Every nation hopes to grow its economy and improve standards of living for citizens. Even if participants fell short of their goal, they likely made an agreement that would move the world closer to a sustainable future.

Climate solutions will need to be achieved by "successive approximation," which means moving closer to a goal incrementally. Barry Richmond's Systems Thinking Skills include "Dynamic Thinking," describing how continuous processes rather than discrete events create the conditions in the real world. Change usually occurs over a period of time rather than immediately. While we may not reach our goal completely, we want to move in the right direction and remember that problem solving often requires an iterative strategy of repeating events and new initiatives.

Bringing the Lesson Home
(continued)

Which parts of the summit groups' presentations were most effective? Why? How could these kinds of learning be applied to general audiences outside of school?

Students will share how they think resistance to climate change initiatives might be overcome so that progress can be made to mitigate the problem. Encourage them to cite previous examples of environmental change and how those situations can inform and inspire us today. While environmental systems are under great strain, progress has been made in many areas. It is important to remain hopeful and positive about the future.

Some examples in the United States:

- *Cars and trucks continue to have improved fuel efficiency.*
- *Since the 1970s, government regulations provide oversight designed to protect the environment.*
- *Recycling rates have increased.*
- *The use of coal has decreased, and renewable energy sources continue to grow as a percentage of electricity generation.*

3.8 Bringing Home Chapter Three

Air

Earth's Atmosphere is a Tiny Ribbon Around Our Planet.

What are some of the major threats to the atmosphere?

Air pollution causes harm to people's respiratory systems. Inhaling particulate matter and dangerous gaseous molecules damages lungs and can introduce harmful chemicals into our bodies. Keeping air clean is not only a problem in urban areas but even for parks and remote territories.

Carbon dioxide (CO_2), which occurs naturally in the atmosphere, has reached dangerous concentrations. Increasing emissions from human burning of fossil fuels combined with deforestation has caused CO_2 levels to rise to the point where climate changes are disrupting ecosystems and human civilization.

Why has it taken so long to implement action steps to reduce air pollution and limit CO_2 and other "greenhouse gas" emissions?

Fossil fuels store very large amounts of potential energy in compact, transportable forms. Modern industrial societies have depended on coal, natural gas, and petroleum for hundreds of years, creating a technological infrastructure that permeates almost all aspects of everyday life. All systems resist change, especially those that support the lifestyle and livelihood of a huge portion of the world's population.

Students will be able to identify factors that work to resist change. Here are some examples:

- *Skepticism and mistrust of science and innovation*
- *Efforts by established corporations and individuals to maintain the status quo*
- *Displacement and disruption caused by change, and fear of the consequences*
- *Economic costs of shifting technologies*
- *Reluctance of leaders to challenge current norms*

How did the systems tools you used in this chapter help explain the environmental systems issues relating to air?

Each of the systems tools helps us understand complex problems.

- *BOTGs show trends and patterns, identifying how variables are changing. Tracking multiple variables can identify causality. Is air pollution getting worse? Are more people becoming sick? What does a graph of CO_2 concentration in the atmosphere show, and how does it relate to average temperatures?*
- *CCs and CLDs can trace feedback loops that lead to environmental problems, such as increasing pollution and rising temperatures. CLDs help us find leverage points, places in a system where policy change can reverse harmful trends.*
- *S/F diagrams and models show the path of "stuff" in a system. When that stuff is carbon in its various forms on Earth or CO_2 in the atmosphere, the diagrams and models help us understand why our climate is changing and what actions are needed to halt that process.*
- *Iceberg diagrams combine tools to explain how system structures have caused the behaviors we can track. The diagrams help us reflect on how our mental models affect our perception of those structures and behaviors.*
- *Simulations like the climate summit show us what is likely to happen without policy changes and how negotiation can bring progress toward a sustainable future.*

What are some of the "big ideas" about atmospheric environmental systems? What can be done on local issues to affect these world-wide problems?

Each lesson highlights an aspect of environmental urgency. A common theme cuts across each topic – air pollution and climate change affect us all. Air pollution diminishes public health, disproportionately affects economically disadvantaged people, and limits economic prosperity in the long term. Climate change acts in similar ways but with longer delays.

Local action multiplied by large numbers of people can make significant inroads against pollution and climate change. These actions can be done in households and communities as well as being instituted as national policies:

- *Shifting away from fossil fuels*
- *Conserving energy*
- *Supporting and promoting policies that favor clean, renewable energy sources*
- *Electing representatives who work toward sustainability*
- *Educating the public to support "green" initiatives*
- *Funding scientific research and development*

How are atmospheric environmental systems connected to land- and water-based systems?

All natural systems are connected and human activities affect them all. Lesson 3.4 explains how carbon exists in different forms everywhere on Earth. This is the case with other molecules as well, and most materials move among land, air, and water systems. Students will have examples of environmental issues demonstrating the connections among different systems. In Chapter 4, students will investigate how oceans are warming and chemically changing due to increased amounts of dissolved carbon.

How do topics in atmospheric environmental systems relate to social justice, public health, and economic prosperity?

Social justice – While everyone suffers from the effects of air pollution, disadvantaged people are harmed to a greater extent than those who can afford buffers. Factories and refineries are built in less affluent areas, and residents often cannot afford air conditioning or filters in their homes. People with less wealth frequently work outdoors and under conditions that expose them to harmful pollution.

Climate change and warming effects usually impact poorer communities and indigenous populations most severely. Storms, floods, drought, melting ice, and other conditions uproot families and disrupt their safe access to food and shelter.

Public health – Air pollution causes or aggravates respiratory illness, especially in children. In some areas of the world, air pollution forces people to stay inside for parts of the year, restricting their ability to benefit from fresh air and exercise. Cities generally have the worst quality of air, but every region on Earth shares the same atmosphere.

Economic prosperity – When air in a city becomes dangerous to breathe, residents, visitors, and workers cannot go about their daily routines. The economy suffers, and productivity declines. The effects of climate change cost enormous sums of money every year. Destruction caused by floods, storms, fires, sea level rise, and other related events harms individuals, businesses, cultural institutions, and governments, necessitating costly repairs and infrastructure changes.

Chapter Four: Water
Living on a Blue Planet
Sustaining our Water Ecosystems

"When the well's dry, we know the worth of water."
Benjamin Franklin

Getting started

Life on Earth is dependent on water. Water covers about 71 percent of the Earth's surface, and fresh water, estimated to be about three percent of the total, is an increasingly precious commodity. Humans depend on water not only for food and drinking supplies but also as a vital part of our climate system. Carbon dioxide concentration, sea level, and temperature have all been rising as climate change becomes an increasingly serious threat.

Students will learn about current threats to Earth's fish stocks and investigate possible strategies to ensure their future viability. They will study the important role oceans play in the complex global climate system.

What will students do and learn in this chapter?

4.1 Protecting Water
Study current threats to water ecosystems and use a variety of systems tools to report on their research.

4.2 Mosquito Nets and Overfishing
Examine the conditions that cause pressure on fish stocks in African lakes. Look at the problem from multiple points of view and learn how "unintended consequences" and the "Fixes that Fail" archetype can impact complex systems.

4.3 Sea Level Change
Perform an experiment to see how melting ice affects sea level. Research how sea levels are changing and diagram the current situation with causal loops.

4.4 Ocean Warming and Chemistry
Experiment to discover how water retains and releases heat. Relate that experience to Earth's oceans and atmosphere. Learn how ocean temperature and chemistry are changing and how that affects short and long-term conditions.

4.5 Running a Fishing Company
Play "Fishbanks," a simulation of the fishing industry. Build a stock/flow diagram or model of the game and consider strategies to ensure a sustainable stock of fish.

4.1 Protecting Water
Clean, Productive, and Free of Plastic

Environmental Systems
What human behaviors are harming Earth's water?

Tools

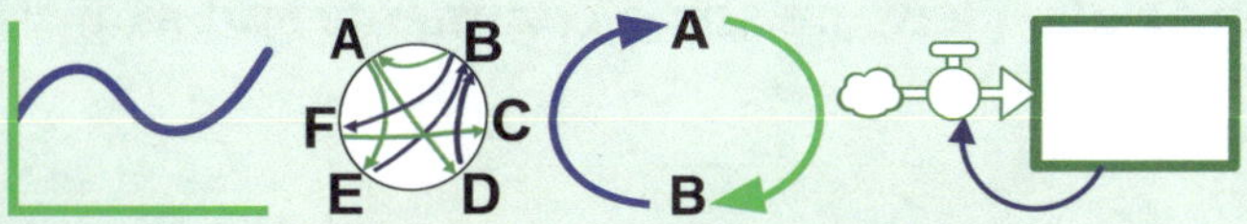

Skills

"... and the great shroud of the sea rolled on as it rolled five thousand years ago."
Herman Melville , Moby-Dick

Overview

About 71% of Earth's surface is covered by water. More than 97% is considered saline, or salt water, with oceans containing about 96.5% of all water. So it's not surprising that the world's precious fresh water supply is overburdened by a growing human population. The oceans are under strain from pollution, overfishing, and climate change, which is altering water chemistry as well as temperature, sea level, and other factors.

Students research threats to Earth's water systems, including the pervasiveness of plastic in ecosystems. They create systems thinking diagrams explaining the problems.

Materials

- Articles or research material about water, p 206
- Resource: Intro to BOTGs, p 215
- Resource: Intro to Connection Circles, p 218
- Resource: Intro to CLDs, p 220
- Resource: Intro to Stock/Flow Diagrams, p 223

Procedure

1. Begin a conversation with students about the importance of water. How do they use water everyday? Where does it come from? Where does it go after use? What current problems are they aware of with water systems?

2. Form pairs or teams of students to investigate the current state of water systems. Use suggested articles linked in the Materials section or have students locate their own research sources. Ask them to report on an important environmental problem with water ecosystems, using at least one systems thinking diagram to help summarize their findings.

   As suggested in previous lessons, choose an interesting format for sharing teams' reports and encourage students' creativity:

- Poster session
- Roundtable discussion
- Presentation modeled on TED talks
- Slide shows or videos
- Print or electronic magazine or newspaper
- Simulated news report
- Report to government panel investigating issue

Ocean Warming and Acidification

These two related threats to water are topics covered in Lesson 4.4. Ocean Warming and Chemistry.

3. Three issues are highlighted in the suggested materials. Here are some comments and sample diagrams:

Drying waterways

Article: "In a Warming West, the Rio Grande is Drying Up" (FIGURES 1 and 2)

A connection circle (FIGURE 1) can be used as a brainstorming tool. Here the thick lines show causal factors that affect the amount of water in the Rio Grande. Thin lines represent the effect of climate change. This is an incomplete connection circle. So far in the diagram, no causal arrows feed back out of "Water in Rio Grande" or into "Climate change."

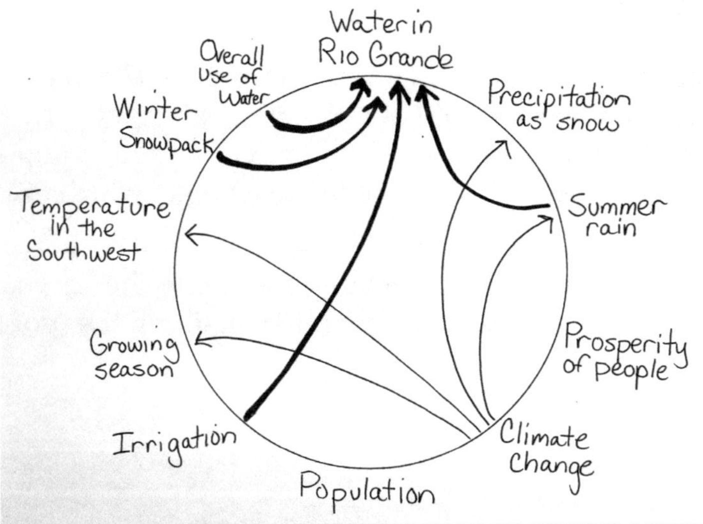

FIGURE 1

Connection circles are an excellent transition and brainstorming tool to elicit feedback loops. Pulling out closed loops will lead to causal loop diagrams. But the example above is incomplete. There is no feedback out of "Water in Rio Grande" or into "Climate change." So constructing a completed causal loop diagram will not be possible (FIGURE 2).

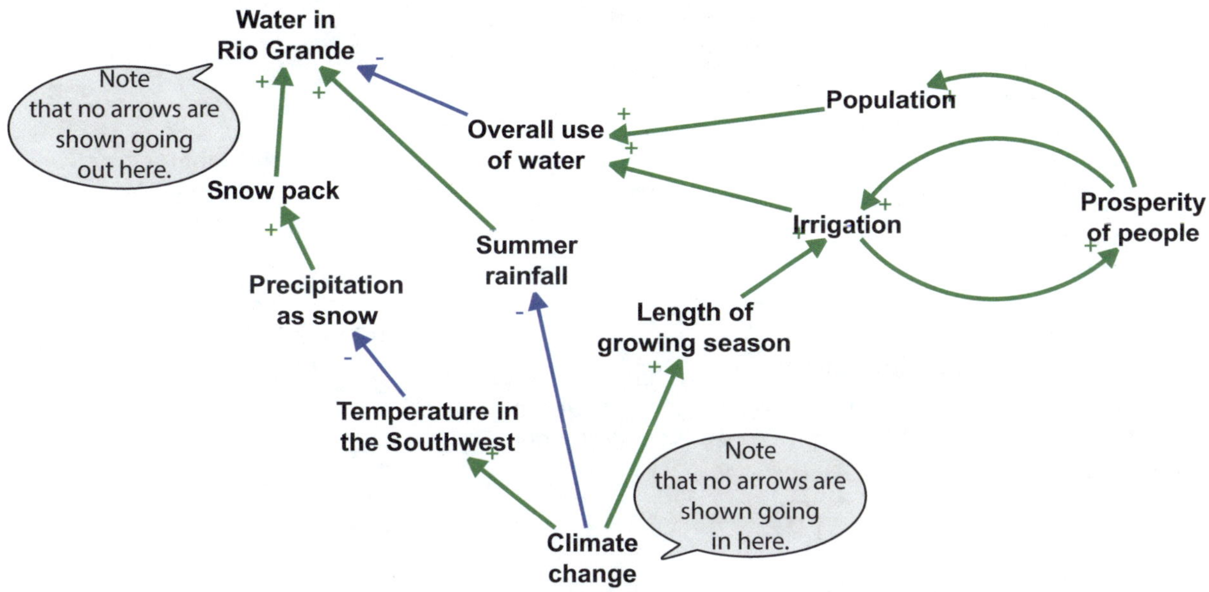

FIGURE 2

To close the loops, the drying up of water in the Rio Grande will have to create motivation for action to mitigate climate change. Student research can find those actions.

Products that pollute

Article: "Is Your Sunscreen Poisoning the Ocean?" (FIGURE 3)

Possible balancing loops (with delays indicated by '=' sign on causal arrow) show how people can become aware of the damaging effects of sunscreen and switch to safer products. Changing behavior to mitigate climate change will have even longer delays. The causal loop here has a dashed feedback arrow to "People visiting reefs." Will tourism need to be limited to save reefs, or can visitors learn to use more care near that ecosystem? "Ecotourism" can be a pathway for people to find ways to help repair and preserve ecosystems through economic support as travelers become educated and informed.

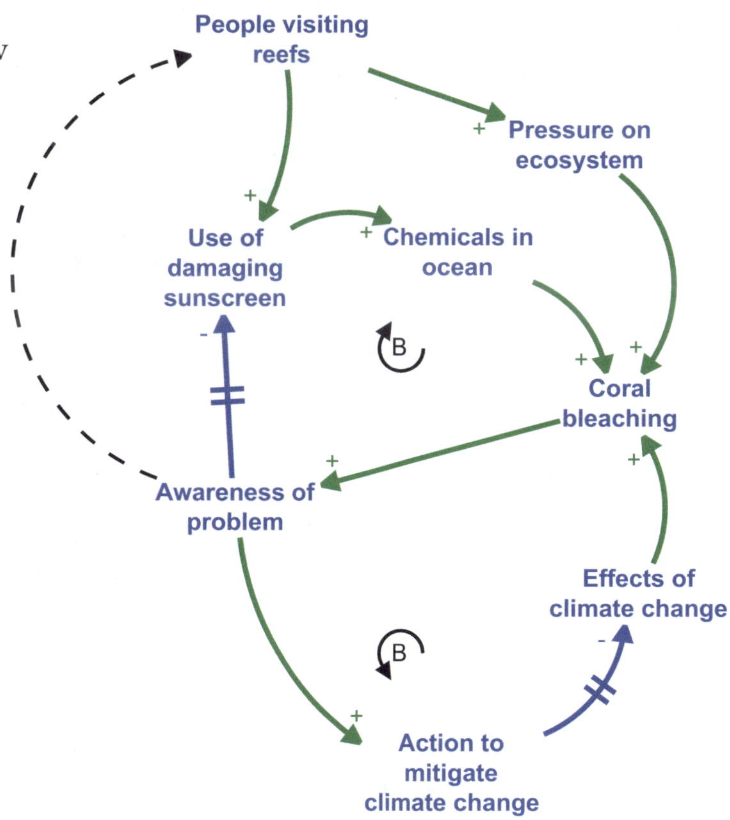

FIGURE 3

Article: "Fertilizers, a Boon to Agriculture, Pose Growing Threat to US Waterways" (FIGURE 4)

A warmer climate causes more rain that washes nitrogen fertilizer into bodies of water. Eutrophication (excessive concentration of nutrients in water) leads to oxygen depletion and harm to aquatic life. Not surprisingly, feedback loops that include solutions also include delays.

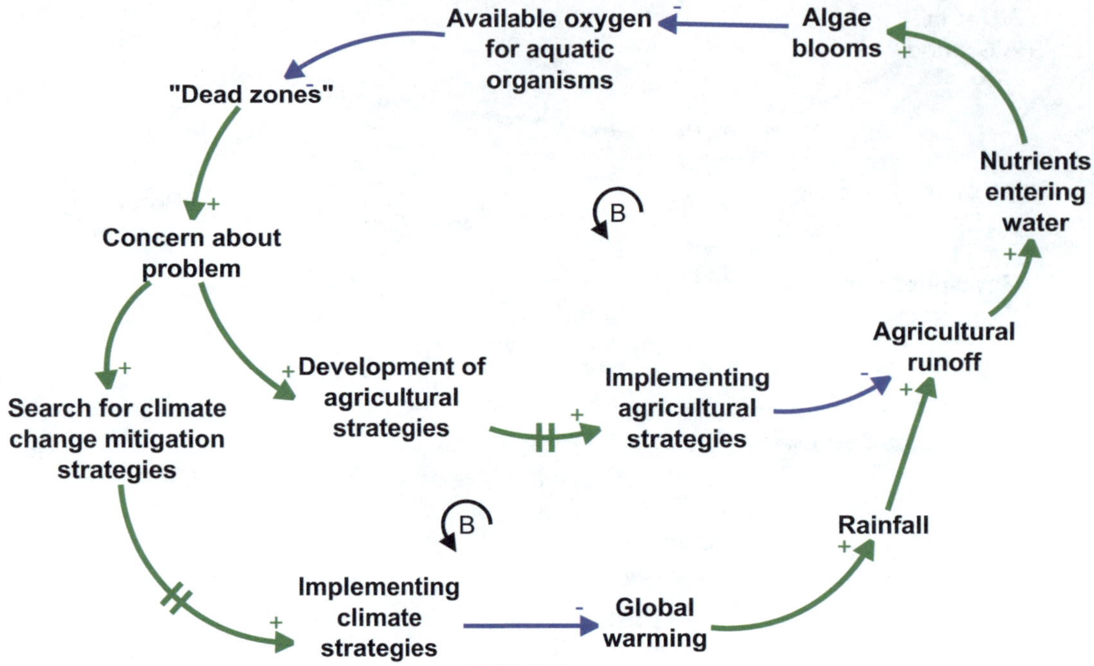

FIGURE 4

Article: "Planning for a Rainy Day"
How urban areas can prevent polluted stormwater from reaching waterways by preserving and creating "green" spaces to absorb and filter it.

Plastics

Article: "The 'Great Pacific Garbage Patch' is Ballooning, 87,000 Tons of Plastic and Counting" (FIGURE 5)
Where is the delay in this balancing loop? As plastic accumulates in oceans and enters the food chain, how long will it take for concern to become action?

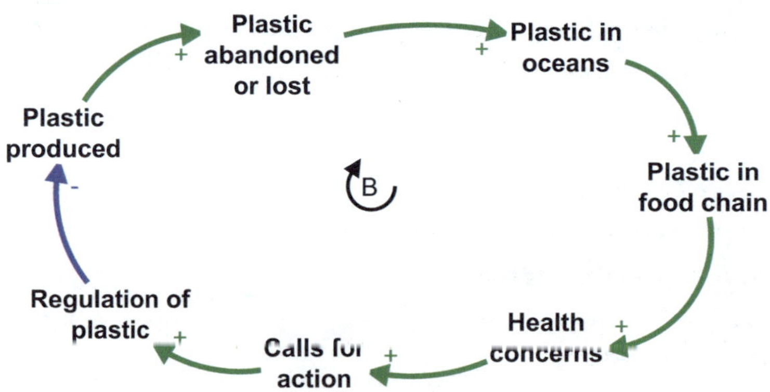

FIGURE 5

Article: "The Immense Eternal Footprint Humanity Leaves on Earth: Plastics" (FIGURE 6)
Have students sketch behavior over time graphs from the data in the article. A stock/flow diagram can show where all that plastic goes.

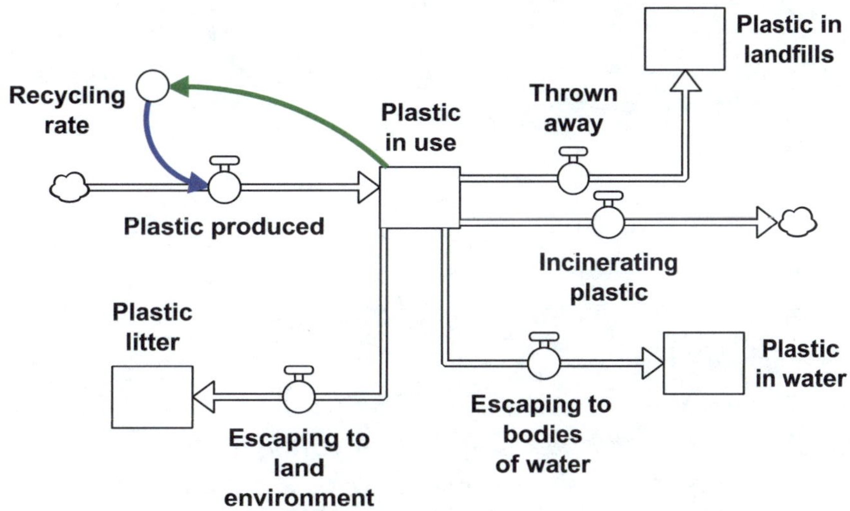

FIGURE 6

Article: "Small Plastic, Big Problem"
Earth's oceans contain disturbing amounts of small pieces of plastic, which enter ecosystems all over the world. "We have a systemic problem," Chelsea Rochman says, "We're mismanaging our waste, and that's coming back to haunt us on our own dinner plates and in our drinking water."

Bringing the Lesson Home

How do systems diagrams help explain threats to water ecosystems?

Each team will find a different way to describe how they used diagrams to explain current problems.

- *Behavior over time graphs record how changes have occurred and what could happen in the future. Water resources are under increasing pressure.*
- *Connection circles encourage brainstorming and help discover the causal links among elements in a system. Feedback loops can be traced and drawn as causal loop diagrams. Water affects and is affected by many factors.*
- *Causal loop diagrams show feedback mechanisms by diagramming the structure and behavior of a system.*

1. *Reinforcing loops drive the elements of a system to continue either growing or decreasing, creating vicious or virtuous cycles, depending on whether the direction is considered a bad or good outcome. Often, human-caused reinforcing loops drive up pollution and overuse resources, severely damaging ecosystems.*
2. *Balancing loops cause elements of a system to oscillate, or change direction as they interact. Nature relies on balancing loops to keep ecosystems stable and healthy, but sometimes human-driven reinforcing loops disrupt them.*
3. *When loops compete in a system, delays occur, and problem solving becomes very challenging.*

Bringing the Lesson Home
(continued)

- *Stock/flow diagrams trace the accumulation and rates of change of the elements in a system. When things move and change, a stock/flow diagram can track their behavior. Following the path of plastic or other pollutants can help find leverage points for cleaning and protecting ecosystems.*

What are common themes that emerge from researching threats to water?

Many water resources are under pressure from a variety of threats. Generally, the problems have been created or made worse by overuse, poor management, pollution, climate change, and other human activities. The solutions have delays and require long-term policy changes that will cause short-term disruptions. The following generic causal loop diagram (FIGURE 7) shows that solutions won't be implemented until the awareness of the problems becomes widely accepted.

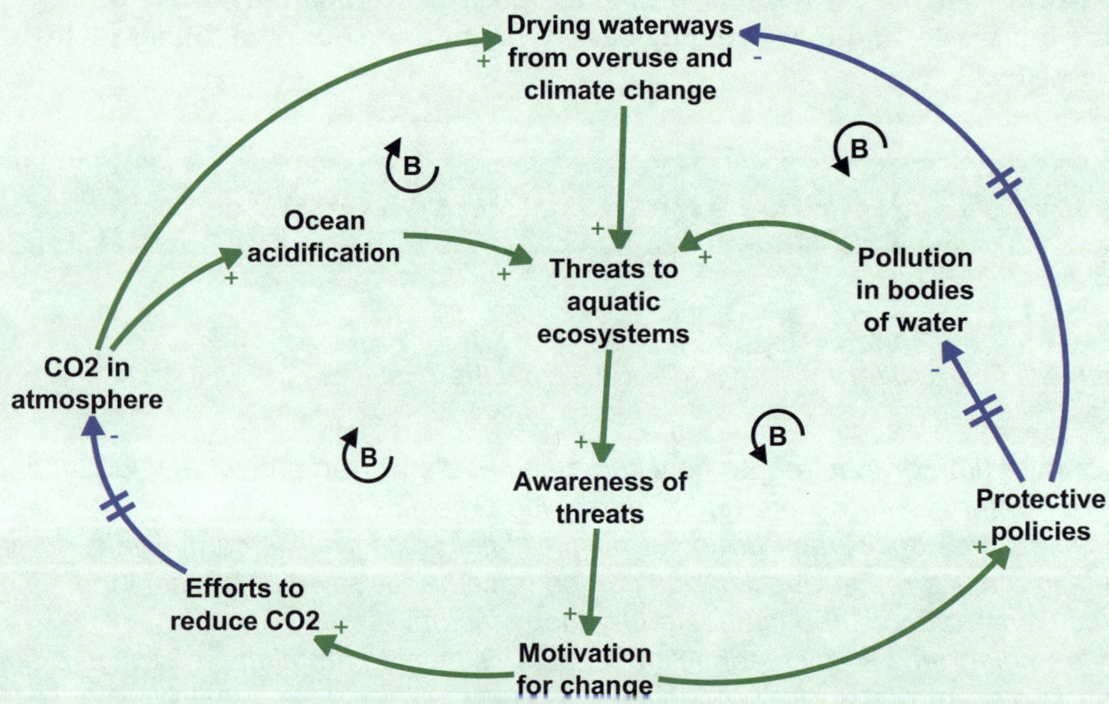

FIGURE 7

After policies are implemented, positive change can take a long time because of delays. Excessive CO_2 and water pollution will persist stubbornly in the environment, even after production is reduced. But there are numerous success stories of pollution reversals leading to recovered ecosystems.

4.2 Mosquito Nets and Overfishing
Unintended Consequences in the Battle Against Malaria

Environmental Systems
How can human behaviors, even those meant to help, harm Earth's ecosystems?

Tools

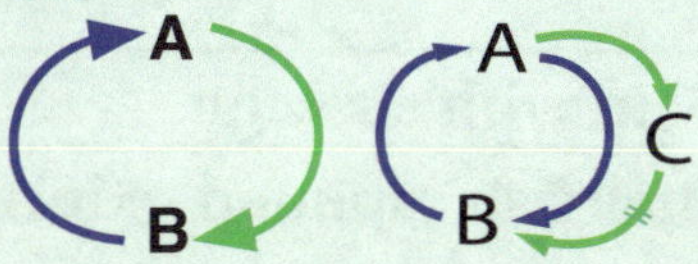

Skill

"There are downsides to everything; there are unintended consequences to everything." Steve Jobs

Overview

Unintended consequences often confound even the most well-meaning problem-solvers. The nets intended to protect people from malaria-carrying mosquitoes ended up causing harm to lake ecosystems in Africa. Systems thinking tools can help us understand the structure of complex problems and enable us to search for fundamental solutions that work.

Students consider different points of view and use systems diagrams to describe the mosquito net problem. They suggest strategies that achieve the desired goals without causing harm.

Materials

- Article and video about malaria and mosquito nets, p 207
- George Richardson's systems diagrams, p 207
- Handout: Stakeholder Concerns, p 150
- Handout: Fixes that Fail Archetype Template, p 234
- Resource: Intro to CLDs, p 220
- Resource: Intro to Systems Archetypes, p 228

Procedure

1. Watch the video and read the article "Meant to Keep Malaria Out, Mosquito Nets Are Used to Haul Fish In."

2. Create teams of students to represent stakeholders in the story. If they participated in "Lesson 2.2 The Tragedy of the Commons," they experienced role-playing people who hold different points of view in an environmental systems problem. Emphasize that each group of stakeholders has one or more overriding concerns, although priorities of people overlap and are not simple. Choose or assign roles including some of the following:

   - people who fish for a living
   - people concerned with avoiding malaria
   - people who struggle to provide food for their family

- international health workers
- environmentalists
- international aid organizers
- journalist reporting on the situation
- government officials

Have teams list what problems their group has and what solutions they have tried. What aspect of their strategy works well, and what drawbacks are they facing? The examples below give one suggestion for each stakeholder. Students should create longer and more detailed lists.

Stakeholder concern	**Successful strategy**	**Problem with strategy**
Catching fish to sell	Using nets – strong, available, effective	Small fish captured reducing replacement stock.
Avoiding contracting malaria	Using nets around beds keeps mosquitoes away from sleepers at night	Nets have some pesticide on them
Feeding family	Nets catch fish effectively	Pesticide enters aquatic ecosystem
Limiting the number of malaria cases	Nets reduce malaria cases	Nets become used for other purposes
Keeping the ecosystem safe and productive	Nets can reduce malaria cases with less pesticide applied	Nets can harm the ecosystem through overfishing and pesticide
Providing aid from charitable contributions	Nets address malaria problems, improving public health	Nets get used for other purposes with consequences that contributors may not approve
Reporting the story	Learning multiple points of view	Problems seem intractable
Governing effectively	Passing and enforcing laws and regulations	Different priorities for groups can lead to problems enforcing laws

3. Project or pass out copies of the template for the archetype "Fixes that Fail;" a simple version is shown in FIGURE 1. Encourage students to work together to see how they might diagram the situation described in the article and video. Use the "Fixes that Fail" archetype as a starting point for causal loop diagrams (CLDs) for each stakeholder group's situation.

   Remember that connection circles can be used as a brainstorming tool for finding feedback loops.

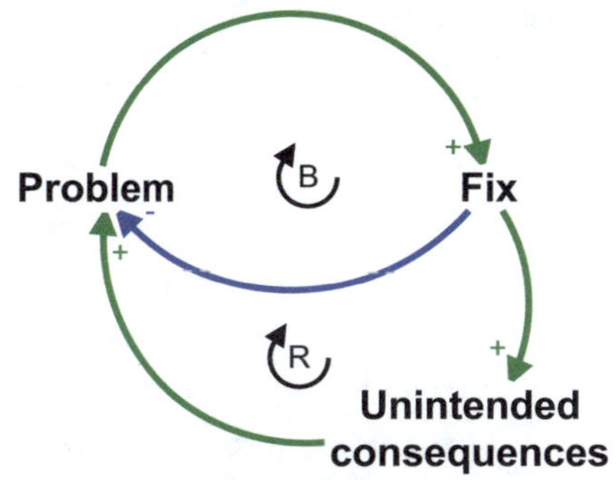

FIGURE 1

4. Share the CLDs when ready. George Richardson, a renowned system dynamicist, created CLDs and a stock/flow diagram for students about the mosquito net situation. Share Professor Richardson's work as an example of how professionals use systems tools to help explain complex problems.

5. Bring students together to brainstorm suggestions for strategies that could achieve the goals of all or most of the stakeholders.

Bringing the Lesson Home

What connections are there between environmental and political problems?

Everyone has basic needs like food and shelter. Feeding one's family is such an important priority that sometimes short-term strategies are necessary, even though long-term consequences will be troubling. The mosquito nets catch fish more effectively than most other methods and provide food immediately. Even if fishermen know that the stock of immature fish needed for future harvesting will be depleted and that the lake can be polluted by the pesticide on the nets, they also know that they have to catch fish to make a living and provide food for their families. The actions and goals of individual people sometimes conflict with environmental and political goals.

Income and resources are not shared equally in most societies. It's easier to be an environmentalist when you have adequate food to eat and a secure home, not to mention income for purchasing goods and services. Governments are responsible for the health and safety of their populations as well as that of the natural environment. In the long-term, people will not flourish if the environment is spoiled.

In prosperous countries, greed often causes overuse and damage to ecosystems. But where people are struggling to survive, just trying to meet their immediate needs can cause harm to the environment in the short-term. Whether a country is rich or poor, the people most harmed by environmental degradation are usually the least wealthy.

"Lesson 5.1 Connecting Systems" expands on the connections among social, economic, health, and environmental problems.

How do the goals of the different groups conflict?

Solutions to a problem can cause other problems, as the mosquito net story shows. Fighting malaria, feeding a population, sustaining a fishing community, … a lot is going on in villages in that region. Each student group can represent the specific needs of their stakeholder group, and they can see which goals overlap and which conflict.

What policy decisions might be made to change the situation?

Some lakes and rivers in the United States and other countries have seen significant environmental improvement compared to past conditions. While it may be more difficult to replicate the successful strategies in developing countries, lessons have been learned. Natural systems can recover if given a chance before it's too late.

Bringing the Lesson Home
(continued)

Students might be able to suggest or research policies and strategies that can feed people, protect them from illness, and keep pollutants out of the environment. Here are five examples:

- *Food aid could be combined with the mosquito net program.*
- *Nets designed for fishing could be produced without pesticide to reduce water pollution and with larger mesh to allow small fish to escape.*
- *Supplemental income aid and food assistance can be made available to people who fish, so overfishing is avoided.*
- *Aid workers could install mosquito nets directly, reducing the temptation to use them for other purposes.*
- *Aquaculture, or fish-farming, could be developed in the region.*

Handout: Stakeholder Concerns

Our stakeholder group is: ____

Stakeholder concern	Successful strategy	Problem with strategy

4.3 Sea Level Change
Rising Waters

Environmental Systems

What is causing sea levels to rise and what are the consequences?

Tools

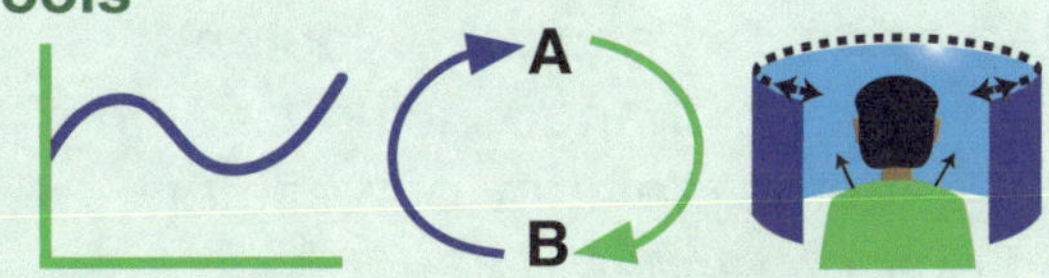

Skill

"Globally, about 145 million people live three feet or less above the current sea level." Jeff Goodell

Overview

As the average temperatures on Earth increase, glaciers melt, and sea levels rise. Less ice means more solar energy is absorbed rather than being reflected, raising temperatures and creating a reinforcing feedback loop.

Students experiment with ice and water to simulate melting glaciers. Then they research sea levels, tracking changes with behavior over time graphs and using causal loop diagrams to explain the causes, problems, and potential strategies.

Materials

- Hands-on experiment materials - ice cubes, transparent cups, rulers or measuring tapes, markers, rocks or other non-absorbent objects to put in cups
- Research material about sea level, p 208
- Background information, p 208
- Articles about sea level rise, p 208
- Multimedia and links to videos, p 208
- Resource: Intro to BOTGs, p 215
- Resource: Intro to CLDs, p 220

Procedure

1. Have each team fill a transparent cup partway with water (FIGURE 1). Have them mark and label the water level. Call this Cup 1. Add two ice cubes to the cup, then measure and mark the new level of the water. How did the water level change? What caused the change?

   (The ice raises the water level because the cubes take up space in the cup and displace some of the water.)

FIGURE 1

2. Fill another cup loosely with rocks. Call that Cup 2. Fill that cup with water so that the level is equal to the first cup (FIGURE 1). The rocks should extend above the water, so add more rocks if the water level covers them. Be sure the water levels in the two cups are equal and mark the level of the water in Cup 2.

   Add two ice cubes to rock-filled Cup 2, being sure the ice cubes rest above the water.

What is the difference between global warming and climate change?

Global warming refers to the measurable increase of the Earth's average temperature over time. Climate change is a term describing changes to Earth's climate that result from the rising temperatures and other factors. Indications of climate change include shifting precipitation patterns, severity of weather events, and other changes.

3. Cup 1 (left cup in FIGURE 1) represents sea ice that will melt in the ocean. Cup 2 (right cup in FIGURE 1) represents ice and snow that will melt on land and enter the oceans. Be sure students predict what they think will happen to the water levels when the ice cubes melt. Leave the cups at room temperature while students begin Step #4.

4. Assign articles to teams or have them begin to research their own sources about sea level rise while the ice cubes melt.

5. After the ice cubes melt, have students measure the water levels in both cups (FIGURE 2). Ask them why the water level rose in Cup 2 but remained nearly constant in Cup 1. Encourage teams to talk among themselves and share their reasoning with the class.

   Example reasoning: The ice in Cup 2 (right) added to the water level when it melted, raising it above the original mark. Although the ice in Cup 1 (left) melted also, the water level didn't change. The ice had been floating in the water, so its volume (i.e., the space it took up) had already contributed to the water level.

   Ask teams to discuss what this experiment represents about the difference between the melting effects of sea ice versus land glaciers.

FIGURE 2

   Example reasoning: When ice that is already in the ocean melts, sea level is unaffected. Melting glaciers and snow packs on land do raise sea level because additional volume is added when they enter the ocean.

   In addition to melting ice and snow from land, other factors contribute to sea level rise, such as the expansion of warming water. The effects can be devastating to communities near the ocean.

6. When each team reports what they learned from their research, be sure they include systems thinking diagrams and draw at least one feedback loop from their reading. Remember that connection circles are a good brainstorming tool for finding causal connections and feedback loops. Trust students to produce meaningful diagrams; some examples are provided in case student groups need guidance.

Albedo and sea ice (FIGURE 3)
Albedo is defined as the proportion of light reflected by a surface. When ice melts and the surface of the ocean has less reflectivity, heat is absorbed and more ice melts, creating a reinforcing feedback loop.

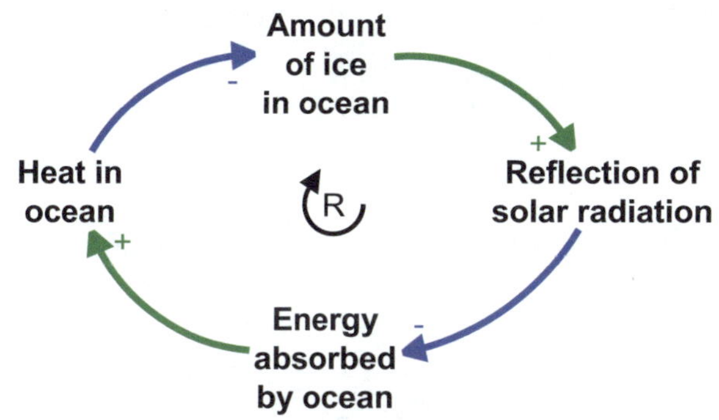

FIGURE 3

Cities and other areas threatened by rising sea level (FIGURE 4)
The dashed connector indicates that moving from "Motivation for action" to "Implementation of mitigation strategies" could be problematic. People resist change and reaching consensus about those strategies will be difficult. The delay mark on the connector between the "Implementation of mitigation strategies" and "Climate change" represents how persistent the effects of CO_2 and other fossil fuel emissions will be.

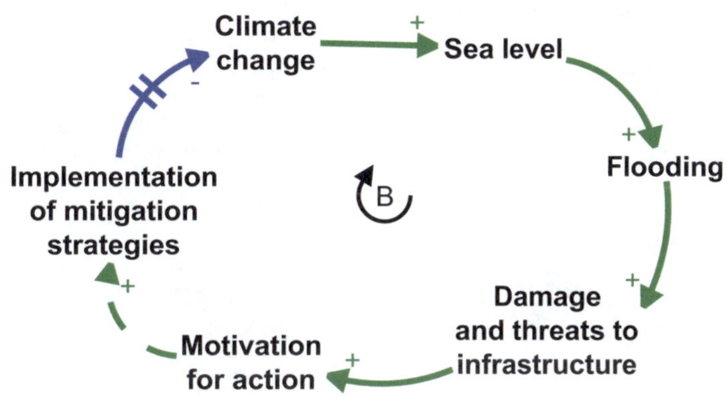

FIGURE 4

Complexity of sea level threats
Specific areas of the world threatened by sea level rise have a variety of problems. The articles students read and research will reveal many factors compounding the difficulty of finding solutions. Systems thinking helps connect factors that contribute to changes, even when those factors may not seem like direct causes. For example:

Article: "Jakarta is sinking so fast, it could end up underwater"
A connection circle of the elements of the problem could include any of the following, some of which may be aggregated into general categories:

- Climate change
- Development
- Natural areas
- Urban planning
- Quality of drinking water
- Sewerage management
- Air pollution from urban sources
- Pollution from burning rain forests
- Trust in government
- Conflict among ethnic groups
- Migration rate from country to city

Students will find causal links among many of these elements.

Bringing the Lesson Home

What does the ice cube activity tell us about how melting ice changes the level of oceans?

Ice that has formed in the ocean doesn't change the sea level much when it melts. But ice and water that flows to the oceans from melting glaciers on land acts like our ice cubes, adding water that raises the sea level.

What other factors contribute to sea level rise?

When oceans warm, the volume of water (the space it takes up) expands. Students should be familiar with thermal expansion; as a refresher, have them think about how liquid-filled thermometers work. When the amount of heat being measured increases, the liquid in the tube of the thermometer expands and rises.

Land-based ice and snow covers much of Antarctica and Greenland. Large sections of those land masses are under huge quantities of frozen water. If students use websites to look at the area of land covered, they may perceive that it does not appear to be shrinking – the land is still covered with ice and snow. But the mass of frozen water is melting and adding steadily to the rise of sea levels.

What are the effects of sea level rise? What can be done to slow the process?

Ocean levels are rising as the overall temperatures on Earth rise. Islands, coastal regions, and low-lying land frequently face flooding. Storm surges threaten broader areas than ever before as warmer air and water fuel increasingly volatile weather. Shorelines erode, and damage to infrastructure causes destruction and economic hardship.

Fossil fuel emissions are trapping more heat in the atmosphere, warming land and oceans on Earth. Until the removal rate of CO_2 *and other greenhouse gases from the atmosphere exceeds the rate at which humans are adding them, average temperatures will continue to rise. Students should be able to brainstorm a list of strategies to combat climate change, including shifting to renewable energy, protecting and expanding forests, and other policies that aim for a sustainable future.*

In the shorter term, mitigation strategies for low-lying regions threatened by flooding include,

- *building or restoring wetland and barrier areas as buffers between the sea and land.*
- *installing floodwalls, gates, storm surge barriers, and levees.*
- *elevating structures and moving their mechanical structures to higher floors.*
- *strengthening building restrictions in flood-prone areas.*
- *replacing paved areas with permeable materials.*
- *creating floating structures that rise with the sea level.*

4.4 Altering the Ocean Temperature and Chemistry

Environmental Systems
What are the major threats to oceans?

Tools

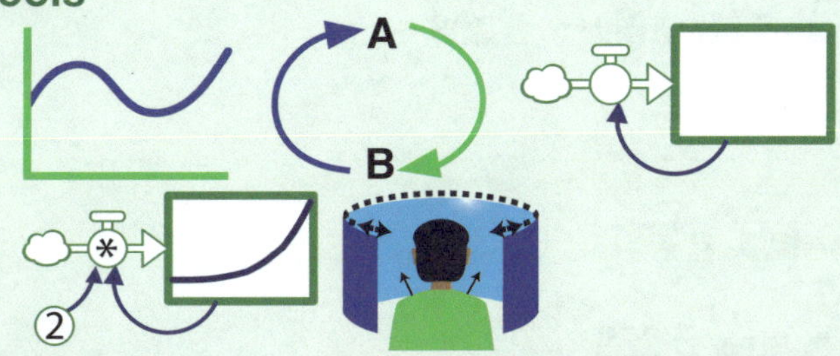

Skill

"We need to respect the oceans and take care of them as if our lives depended on it. Because they do." Sylvia Earle

Overview

Scientific consensus predicts significant climate change as a result of increasing CO_2 levels in the atmosphere. Effects of the warming temperatures have been recorded and studied since the late 20th Century. Ocean temperatures and chemistry are changing also. Water heats up and cools down more slowly than air, making oceans an important regulator of climate.

Oceans cover approximately 71% of Earth's surface. They have a profound role in determining our planet's varied climate regions, interacting with the atmosphere in a complex system that contains delays. In his list of the characteristics of complex systems, Professor Jay Forrester included "Cause and effect are not closely related in time or space."

Students investigate current conditions and possible future outcomes as fossil fuels are burned. Higher CO_2 concentration in the ocean lowers the pH of the water, and creates a more acidic environment that threatens the health of aquatic ecosystems.

In this three-part activity, students:

1. study how water retains and releases heat, working with a lab exercise and a computer simulation;
2. sketch graphs to see the delays among length of daylight, air temperature, and ocean temperature; and
3. research how ocean temperature and chemistry affect both short and long-term conditions, including how increased CO_2 levels are changing ocean chemistry.

Materials

- Lab Materials
  - ◊ Device to boil water
  - ◊ Heat-resistant cups or containers

  - ◊ Lab thermometers or computer-coupled temperature sensors
  - ◊ Timers or clocks

- Markers and large poster paper
- Diagramming and modeling materials
  - ◊ For diagramming, no materials necessary; diagrams may be drawn by hand.
  - ◊ For computer drawing and model building, computers with system dynamics software

- Lesson 4 from *The Shape of Change*, "It's Cool," page 208
- Information about length of day and temperature, p 208
- Suggested articles and resources about ocean heat and chemistry, p 208
- Resource: Intro to BOTGs, p 215
- Resource: Intro to CLDs, p 220
- Resource: Intro to Stock/Flow Diagrams, p 223
- Resource: Intro to System Dynamics Computer Models, p 224
- Resource: Intro to Systems Simulation, p 225
- Resource: Intro to System Dynamics Modeling Software, p 238

Procedure

Part One: Cooling water experiment

1. Follow the procedures for Lesson 4, "It's Cool," from *The Shape of Change*. This activity is a lab experience about energy transfer and the rate of cooling using very hot water. Be sure to follow safety procedures and have adequate adult supervision. If it is not possible to have students perform the lab, you may present it as a demonstration and possibly use a computer simulation.

   Use the debriefing questions in the "It's Cool" lesson plan to reinforce the lab experience.

   Construct a stock/flow diagram as described in the lesson. You may continue on to build a computer simulation.

2. Have students relate the cooling curve from the experiment to heat in the ocean. When water is warmer than the surrounding air, heat escapes from the water over time. When the air is warmer than the water, the opposite happens. An enormous body of water like an ocean will take a long time to cool or warm. The USGS article, "Heat Capacity and Water," contains a clear summary of why oceans warm and cool more slowly than air.

   NOTE - Be sure students know that air is not heated directly from solar radiation but by heat radiating from the Earth's surface.

Part Two: Graphing to see delays

3. Heat energy from the sun warms Earth and then a portion of the energy radiates back toward space, causing air temperature to change over the course of a year. Because of the tilt of the Earth and its path around the sun, the amount of heat energy on any given area varies throughout the year. What is the relationship between the hours of

Emphasize the shape of change

Remind students that their BOTGs should show the overall curve of the data. It isn't necessary to plot 365 points. They can find the length of day for one day per month, or plot the solstice and equinox points, and draw a smooth curve connecting those points. Just plotting a few points and connecting them with a smooth curve saves time and emphasizes the shape of the graph. Students will have a sense of how the hours of daylight oscillate over time, and how the height of the curve correlates to the latitude of the location.

daylight and the average temperature of the air? Do air and ocean temperatures rise and fall together?

Have teams of students draw a behavior over time graph (BOTG) of the amount of daylight per day for the location of their school over a period of a year. Graph and compare a place at a different latitude. They can find data at a variety of websites.

Encourage students to sketch their graphs by hand. Some teachers may choose to have students import data into a spreadsheet and create a graph electronically. While that's a valuable skill to practice and a valid pedagogical strategy, drawing a graph by hand can reinforce systems thinking skills, especially looking at the "big picture" and tracking how variables change over time.

4. Instruct students to add a line showing average daily temperature on the same BOTG, making a prediction of its shape before finding the data. The temperature scale may be written in a different color or drawn on the right side of the graph, creating an additional vertical axis. Find data at websites linked in Appendix B.

   Again, instruct students to find a few points and sketch the graph as a smooth curve. They should note that the temperature graph has a similar shape to the length of day but is shifted about one month later in the year.

5. Have those students who are graphing temperature data for coastal locations find a third variable to graph – average ocean temperature. Notice the differences between this graph and the other two.

   Here's a sketch of a three-variable graph (FIGURE 1).

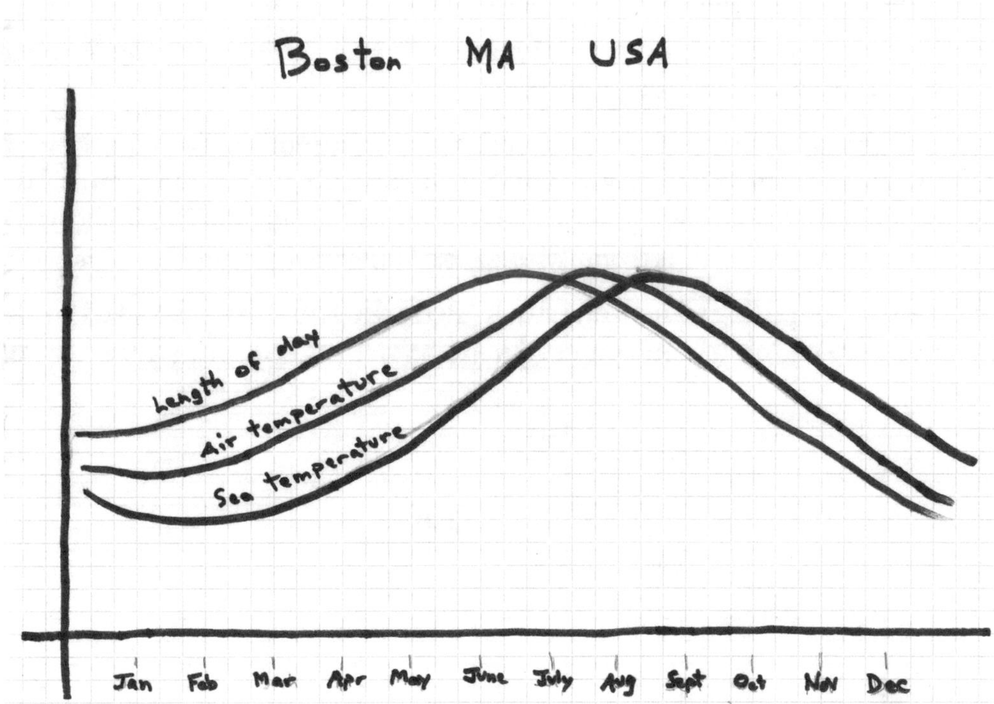

FIGURE 1

Notice there are no scales on the y-axis. The curves show when the average maximum and minimum values occur, emphasizing the delays between the length of day and temperatures. Students can plot graphs more specifically and with carefully drawn scales, but that will extend the time necessary to complete the work.

6. Compare the length of day graphs to the temperature graphs. If teams have chosen different locations to graph, share and compare data. What is the relationship between the hours of daylight and the temperature? How do the air and sea temperatures compare? Why are there delays in the temperature curves following the shape of their accompanying hours of daylight curve? The prompts for analyzing BOTGs can be useful here for either oral or written presentations:

   - What aspects of the data are changing?
   - How are things changing?
   - Why are the changes happening?

7. Be sure some teams of students research the effect of oceans upon the air temperatures at various locations and latitudes. Are coastal locations different from inland places, even at the same latitude? When sharing their findings, students should include connection circles and feedback loops to illustrate the connections they can trace between ocean temperature and climate.

   An online search will find many valuable websites. The United States government provides useful starting points, including the NOAA website.

Part Three: Ocean warming and chemistry

8. Students are probably familiar with the effect of rising CO_2 levels in the atmosphere. The greenhouse effect is well known and frequently cited in science reports and by the news media. It is the subject of Lesson 3.5 The Greenhouse Effect.

   Less well known is that increased CO_2 in the atmosphere is leading to more CO_2 in the oceans. What does this mean for the health of Earth's oceans?

   Assign student teams to research ocean warming and CO_2 levels. Use the suggested articles or have student teams research on their own.

   Share teams' findings with presentations based on their research. They may reference the carbon cycle in their reports. (See Lesson 3.4 The Carbon Cycle for diagram.) Encourage them to use any of the systems tools they prefer: behavior over time graphs, iceberg diagrams, connection circles, causal loop diagrams, and stock/flow diagrams.

   Examples of a causal loop diagram and a stock/flow diagram are provided in the Bringing the Lesson Home debrief section for guidance. As always, trust student teams to create their own diagrams.

Bringing the Lesson Home

How do ocean temperatures affect overall global warming?

Oceans cover a large portion of the surface of the Earth and therefore receive a lot of radiant energy from the sun. Oceans store heat energy. Because water cools more slowly than air, oceans generally stay warmer than land during winter. Warmer oceans affect climate in several ways, including air temperature and amount of precipitation over many regions of the Earth.

What are the major differences among the graphs of length of day, air temperatures, and ocean water temperatures? How can you explain these differences?

In the northern hemisphere, the date with the most daylight occurs around June 21. The average date when the highest temperature occurs is later, happening about one month later in mid to late July. This delay is due to the Earth heating up from solar radiation and gradually warming the atmosphere. The highest ocean temperature happens even later in the year, as the sea water takes longer to warm up. The ocean retains heat longer than air and cools more slowly, staying relatively warm in the fall compared to falling air temperatures. So while the three graphs have similar shapes, their peaks and valleys do not line up because of those delays.

What happens when CO_2 levels in the ocean rise?

CO_2 dissolves in water. A chemical reaction occurs that forms carbonic acid. The pH level of the ocean water drops, and less calcium is available for creatures to form shells and build coral reefs.

How can you represent the process of ocean acidification using systems thinking diagrams?

Students should be able to find behavior over time graphs during their research. They will also create a variety of diagrams that reflect their thinking. This causal loop specifically looks at coral bleaching (FIGURE 2).

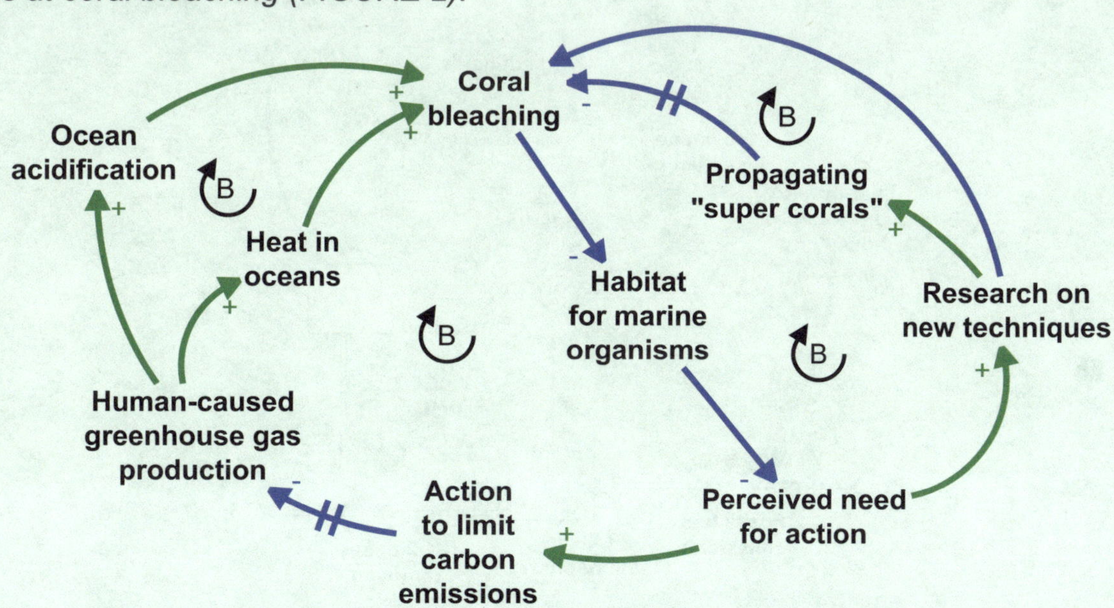

FIGURE 2

Bringing the Lesson Home
(continued)

The following balancing causal loop explains that CO_2 increases ocean acidity and suppresses growth of creatures, affecting food webs. As public awareness grows, initiatives should be implemented to reduce atmospheric carbon, albeit with delays (FIGURE 3).

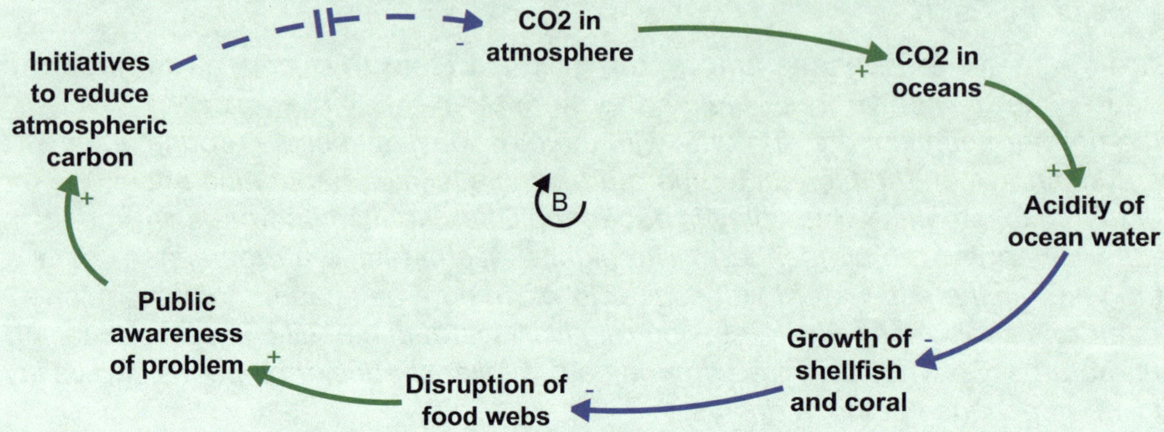

FIGURE 3

Here is a simple stock/flow diagram of the same concept (FIGURE 4).

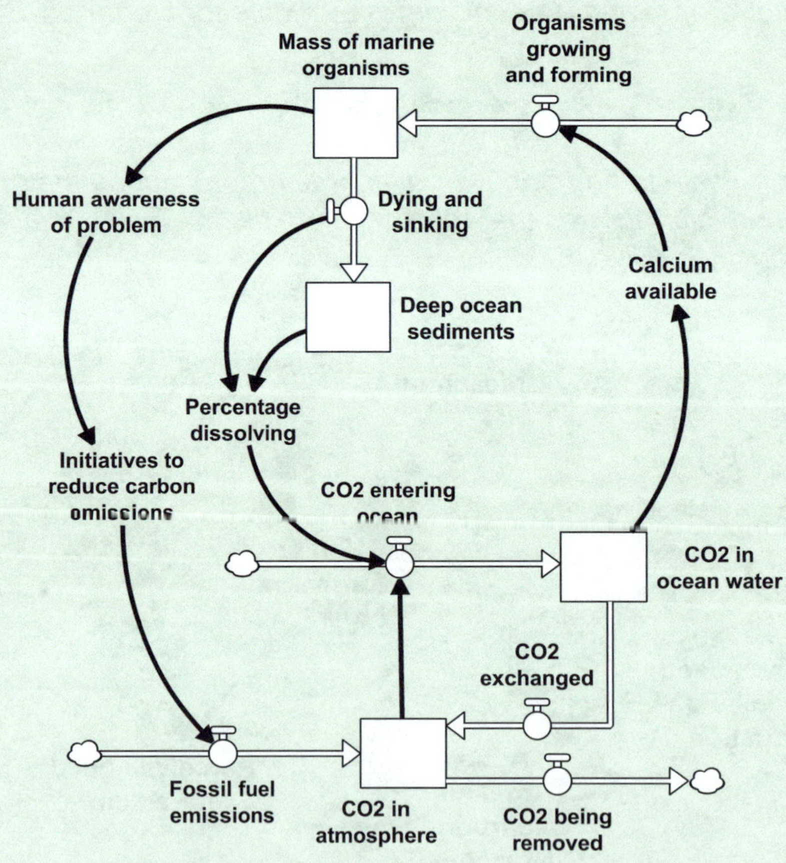

FIGURE 4

4.5 Running a Fishing Company
Sustainable Resources From the Ocean

Environmental Systems
What are the challenges of maintaining a sustainable fishing industry?

Tools

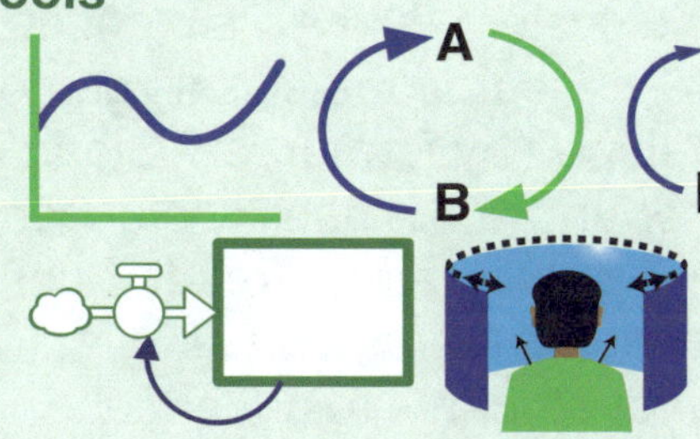

Skill

"Give a man a fish and you feed him for a day; teach a man to fish and you feed him for a lifetime."

Maimonides

Overview

Overfishing, along with pollution, climate change, and other factors, has reduced ocean fish stocks significantly over the past two centuries. Because much of the world's population depends upon the oceans for food, the decline in available resources poses serious threats to a sustainable future.

Students play the simulation game, "Fishbanks." They experience what happens when companies compete for a limited resource. After they have played and debriefed the game, students create a stock/flow diagram representing the structure of the Fishbanks simulation.

Materials

- Online Fishbanks simulation, p 209
- Computers with Internet access
- Handout: Fishbanks packet from Creative Learning Exchange, p 209
- Reference article, "Building the Fish Banks Model and Renewable Resource Depletion," p 210
- Resource: Intro to BOTGs, p 215
- Resource: Intro to CLDs, p 220
- Resource: Intro to Stock/Flow Diagrams, p 223
- Resource: Intro to Archetypes, p 228
- Resource: Intro to Simulation, p 225

Procedure

Running the simulation

1. Familiarize yourself with the materials from the Creative Learning Exchange and register as an educator through the MIT Sloan School. If you are planning to run the simulation with more than a few teams, consider inviting other adults to assist you.
2. Form teams of students and explain that they are going to play a game in which each team operates a fishing company.
3. Follow the procedures outlined in the simulation materials. Be sure to schedule 1-2 hours for a long enough simulation. Debriefing materials are included. Additional debriefing questions are included in the Bringing the Lesson Home section at the end of this lesson plan.

Building stock/flow diagrams after the simulation

4. Depending on students' experience with stock/flow diagrams, choose whether they work in their teams or as a whole class.

   If you are going to conclude the lesson with stock/flow diagrams rather than working models, it isn't necessary to have one consensus diagram. Each team should create a diagram that tells the story of the Fishbanks game the way they interpret it.

   This clear and concise causal loop diagram is from the Creative Learning Exchange materials (FIGURE 1).

Description of the Fishbanks simulation from the MIT Sloan School website

Balancing Sustainability and Profitability

Fishbanks is an interactive, web-based simulator developed by MIT Sloan researcher and system dynamics pioneer John Sterman, based on the classic game originally developed by former MIT Sloan faculty member and sustainability pioneer Dennis Meadows. Offered for free through MIT Sloan School of Management, it is a unique way for students to understand the challenges inherent in sustainably managing finite resources.

Players in Fishbanks are divided into teams of fishing companies. Each team seeks to maximize its profit by competing to catch as many fish as possible without depleting stocks. The oceans are home to a pool of fish that is accessible to each fleet. The object and challenge of the game is to catch as many fish as possible and not overtax fish stocks while in direct competition with the other fleets.

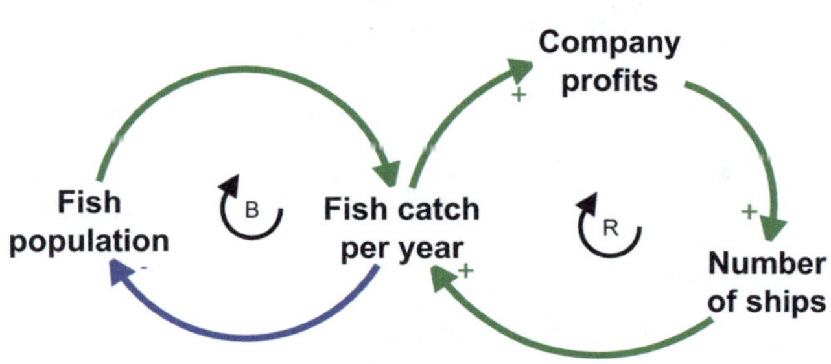

FIGURE 1

Causal loops can be good starting points for building stock/flow diagrams.

Simple four-step guide to building stock/flow diagrams including an example three-stock model with "Money in bank" as a stock (FIGURE **2).**

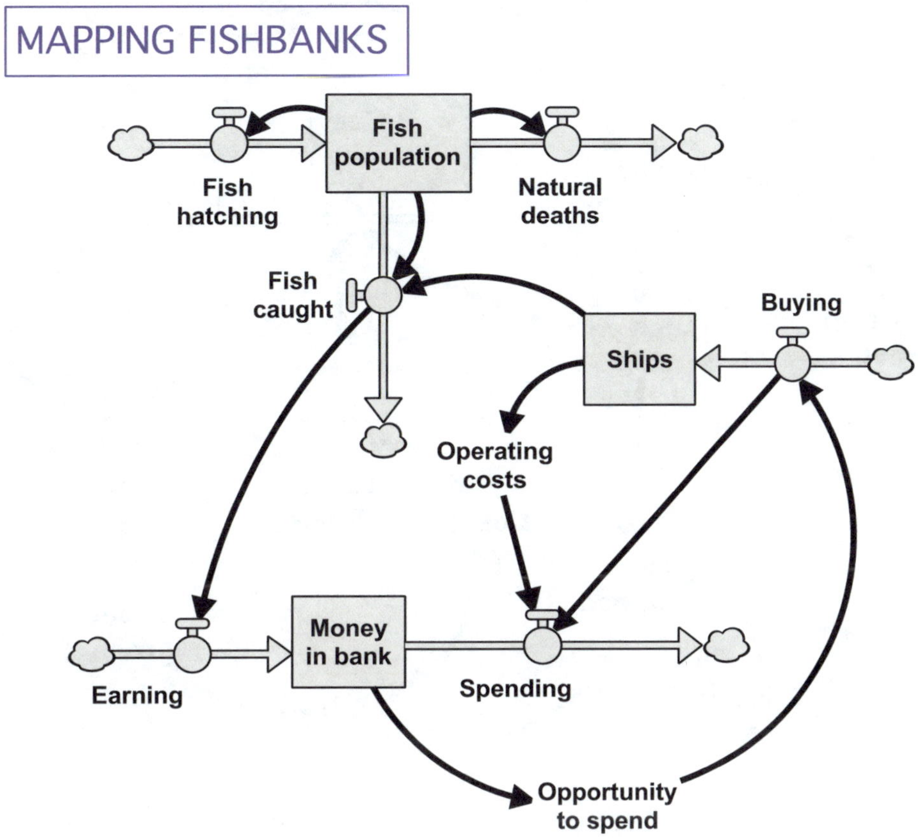

1. Find stocks - try to limit how many.
2. Add flows - Only add flows that add to or subtract from stocks.
3. Draw connectors - they CANNOT go into stocks.
4. Trace loops.

FIGURE 2

The following stock/flow diagram has two stocks. Students might be able to expand the diagram with more connections (FIGURE 3).

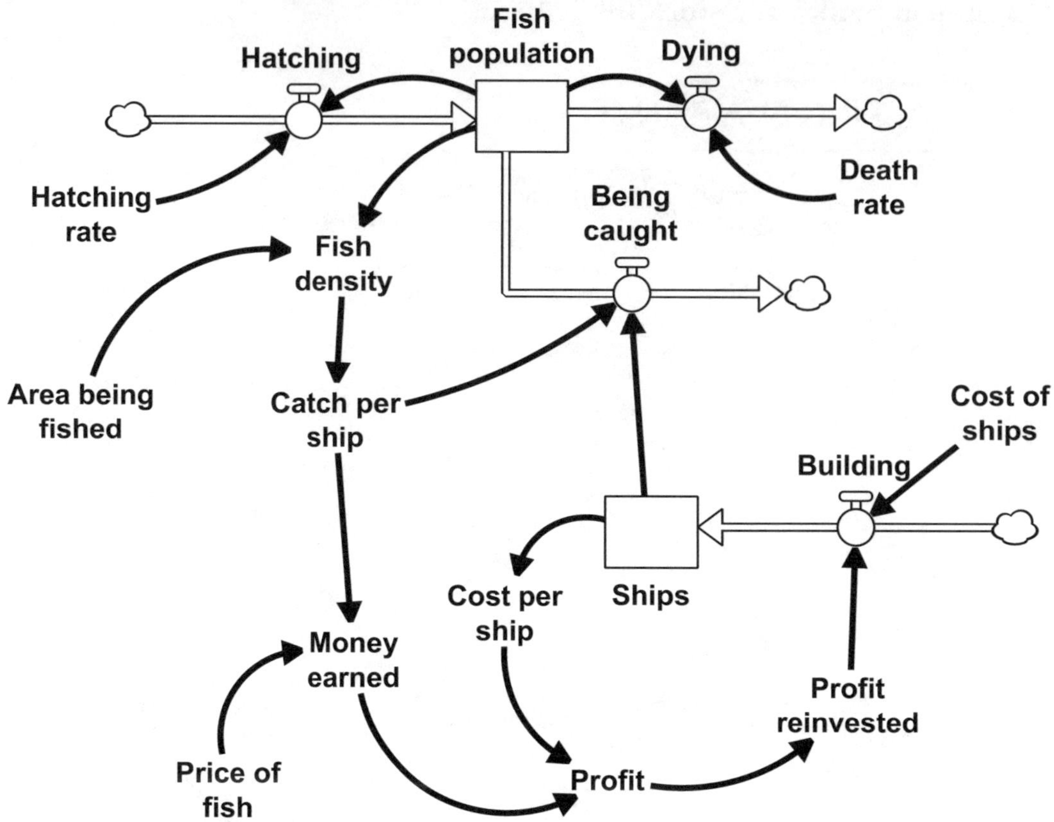

FIGURE 3

For example, what affects the "Area being fished?" What changes the "Price of fish?"

5. For classes continuing onto model building, refer to the MIT reference article link. Some of the language may seem technical to inexperienced modelers but equations are given for their two-stock model. Instructions are provided for building the system dynamics model as well as worksheets with instructions suggesting how to modify the model to test a variety of policies to avoid overfishing.

6. Students can share their stock/flow diagrams as part of the debrief section below, "Bringing the Lesson Home." What ideas do they have for making the fishing industry sustainable?

Bringing the Lesson Home

What is the object of the game?

Players compete to make money. In order to make money, each team tries to catch as many fish as possible. But even the teams who make a lot of money early in the game end up with very few fish to catch. So the goal of catching a lot of fish helps to put your own team out of business. The true object of the game might be to figure out how to maintain a sustainable fishing industry.

How is this an example of a systems archetype?

The fish stocks are depleted because the ocean is regarded as a common resource available to all. When teams attempt to maximize their catch, the fish population crashes and eventually no one can benefit. This type of situation is an example of the Tragedy of the Commons archetype (Appendix C).

How is this environmental systems problem similar to others?

The archetype called "Tragedy of the Commons" can occur when a shared resource is overused. [Lesson 2.2 The Tragedy of the Commons - How Should Public Land Be Used? introduced this archetype and others.] The fish in the ocean are seen as not belonging to anyone until they are caught, so some fishing companies continue to deplete the stock at a rate greater than it can replenish itself. Water, minerals, forests, and other natural commodities are all subject to decline or pollution if too many companies overuse them.

How did tracing some of the feedback loops explain the outcome of the game?

Every team tried to make more money. To make more money, they sent more ships out to catch fish. As the fish population went down, it took more ships to catch the same number of fish. The fish population went down and the number of ships kept going up.

How could you change the outcome of the game? What strategies did your team suggest to prevent overfishing and collapse of the industry? How might they work? What could be the effect of your policies on the communities involved, both long and short term?

Teams will have suggested a variety of strategies. These may include:

- *limiting the number of ships*
- *restricting the use of technology*
- *regulating the fish catch per ship*
- *imposing taxes on ships and/or fish caught*
- *establishing "no fishing" zones as sanctuaries for fish populations*
- *increasing "fish farming" to supplement wild-caught species*

You may repeat the simulation with students altering their behavior. With teams cooperating rather than competing, can the outcome of the game be changed? Can people create a sustainable fishing industry and still support their families?

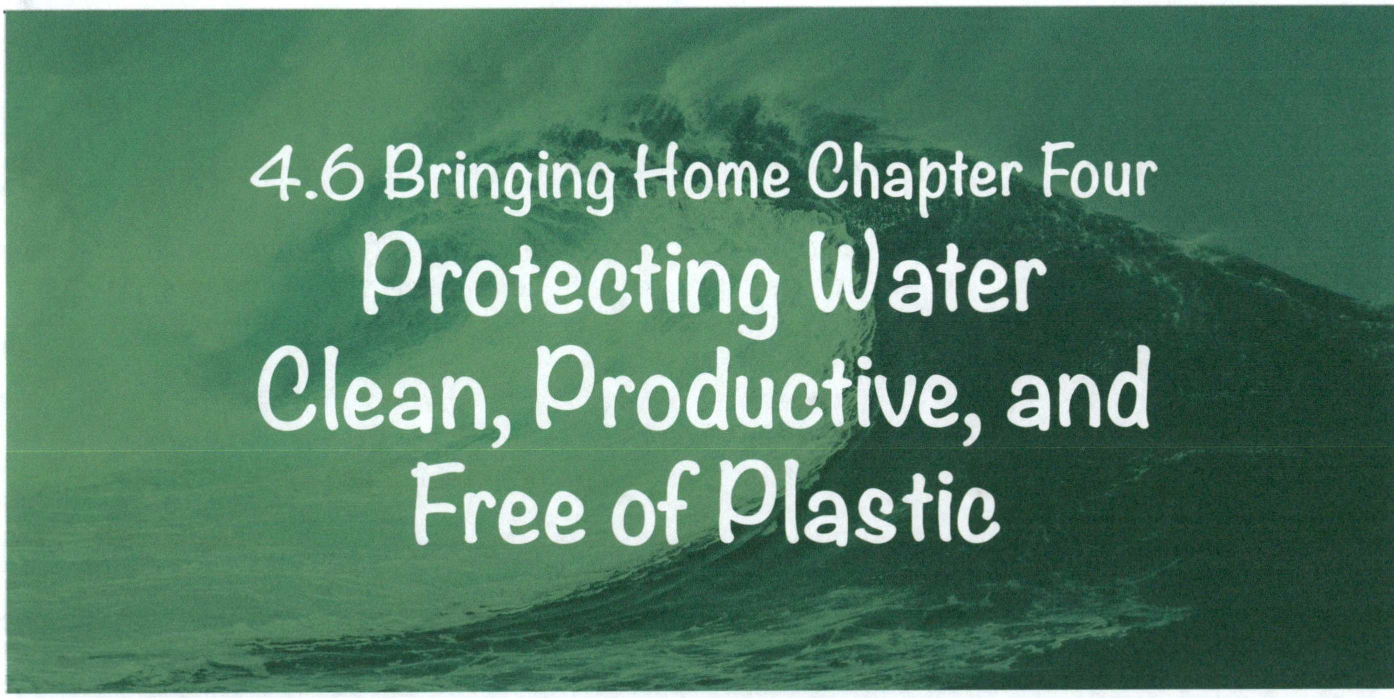

4.6 Bringing Home Chapter Four
Protecting Water Clean, Productive, and Free of Plastic

What are some of the themes that make water-based, environmental systems critically important to a sustainable future?

Water is a vital commodity for life on Earth. As human population has grown and developed technology to expand and flourish, water resources have come under increasing pressure. Chapter 4 highlights some of the issues threatening water ecosystems:

- *Pollution – plastic, chemicals, run-off from agricultural sites, and other waste products entering water systems degrade their quality and can render them unfit for use and animal habitation.*
- *Supply – overuse by humans continues to exhaust aquifers, rivers, and other sources of fresh water.*
- *Unsustainable harvesting of resources – many fish species cannot regenerate to match the rate at which they are being caught.*
- *Changes in oceans – the temperature, chemistry, and level of the world's oceans are changing due to pollution, climate change, overuse, and other human behaviors.*

How do water systems interact with land and atmospheric systems?

Students should be familiar with the water cycle, the process by which water moves throughout the world's land, air, and water systems. Water stores much of the heat energy the Earth receives, and the world climate cycles and seasons depend on the oceans for stability and predictability. Oceans function as a "carbon sink," absorbing and storing carbon from the air and land.

All bodies of water receive material from land. Deliberately or not, human debris is often carried away or dumped into water. Chemical concentration increases over time, resulting in dangerous levels of toxicity.

Because water moves all around the Earth and is necessary to supporting life, sustainability depends on healthy water-based environments.

How did systems thinking tools and principles help describe and explain water-based environmental systems issues?

- *BOTGs – Water systems often have delays, so it's important to look at longer-term changes.*
- *Connection circles and CLDs – Feedback loops trace the causality of deteriorating conditions and identify where possible strategies can be implemented. Motivating people to change often depends on raising awareness of problems and suggesting paths for correcting them.*
- *S/F diagrams and models – The dynamic stocks of plastic, carbon, and fish populations in oceans present difficult challenges in environmental science. S/F thinking focuses strategic planning on inflows and outflows, helping to determine action plans for the future.*
- *Simulation – The Fishbanks simulation demonstrates how a Tragedy of the Commons can reduce natural resources below sustainable levels and encourages consideration of policies for change.*

How do topics in water-based environmental systems relate to social justice, public health, and economic prosperity?

Social justice - As with all environmental issues, water problems affect disadvantaged people most severely. Clean water for drinking, cooking, and bathing should be a human right but many communities lack safe supplies. Sources of pollution are frequently located near these communities, and oversight can be lax.

Public health – The pollution of water sources causes serious health consequences. Much of the world population's food sources are threatened by water pollution. Climate-related changes harm agricultural production and are forcing large numbers of people from their homes, creating crises of climate migration in many areas of the world.

Economic prosperity – Without adequate clean water supplies, societies cannot flourish. If industries pollute the water supply, eventually the health consequences will damage the lives of people they need to produce and buy their goods. For example, in the 21st Century, the use of hydraulic fracturing, or fracking, to extract petroleum and natural gas became widespread. Critics cited pollution of household water sources as one of several objections to the practice, and understanding the true cost of fracking began to turn public opinion against it.

Chapter Five puts environmental systems together with other systems – racial justice, economic activity, and public health. Students find the connections among these systems that create the structures driving the behaviors we see everyday. They choose topics to research and report upon effects to them and their communities.

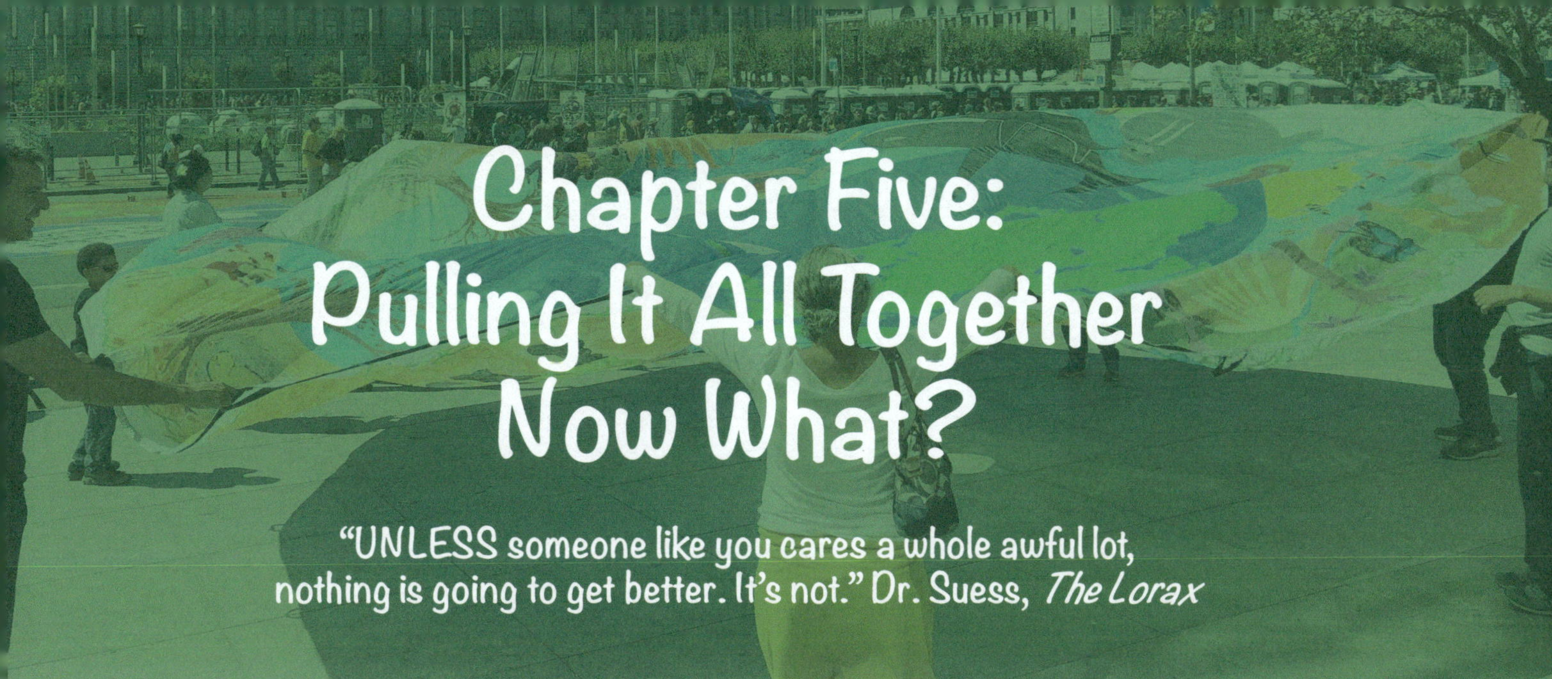

Chapter Five: Pulling It All Together Now What?

"UNLESS someone like you cares a whole awful lot, nothing is going to get better. It's not." Dr. Suess, *The Lorax*

Getting started

Sometimes educators and parents worry that students will become pessimistic and lose hope when they learn about negative events in the past and the difficult problems of today. We have an obligation to help them discover what they can do, so they assume responsibility for the world and strive to make it a better place.

Barry Richmond used the phrase 'systems citizens' to describe how students (and others) can grow into responsible members of their communities and inhabitants of Planet Earth. Systems citizens study and practice the systems thinking skills Barry Richmond wrote about eloquently, and work to bring about positive change. They strive to see things from multiple perspectives and understand how actions affect different people and the world around them. Having students investigate topics they really care about will help them think about their roles as systems citizens.

One principle, often attributed to the philosophy of the Iroquois Nation, is "In every deliberation, we must consider the impact on the seventh generation …."

Delays are common to all systems, and every generation deals with actions and decisions made long ago. Change can be slow and often difficult. Another wise statement, this one a Chinese proverb, gives us encouragement.

"The best time to plant a tree was 20 years ago. The second best time is now."

What will students do and learn in this chapter?

5.1 Connecting Systems

Explore the connectedness of environmental systems with other issues in society – social justice, public health, and economic prosperity. Use systems tools to look for leverage points and strategies that will create a more just, sustainable, and prosperous society while protecting and regenerating the environment.

5.2 Choosing a Topic

Choose an area of environmental systems to explore in depth – extend a topic studied earlier, continue connections from Lesson 5.1, or research an issue with local significance. Report on their research, choosing systems tools and other media to educate and inform their audience.

5.1 Connecting Systems
Environmental Systems, Social Justice, Public Health, and Economic Prosperity

Environmental Systems
How are issues of environmental systems, social justice, public health, and economic prosperity interconnected?

Tools

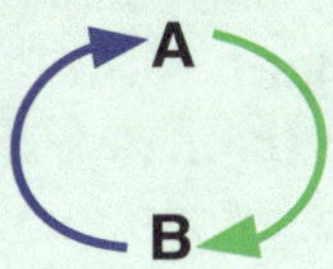

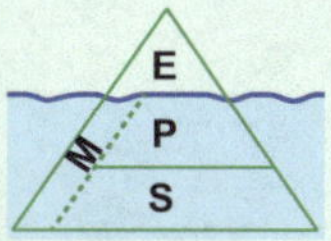

Skills

"Saving our planet, lifting people out of poverty, advancing economic growth... these are one and the same fight."
Ban Ki-moon

Overview

Environmental systems support all life on Earth, and they affect all aspects of our civilizations. The fair and equal treatment of people (social justice systems), their physical and mental well-being (public health systems), and their access to goods and services (economic system), all depend upon the quality, sustainability, and regenerative capacity of environmental systems. The four systems are connected by feedback loops and form an intricate web. Maintaining healthy environmental systems is the right thing to do, not only because humans have become the stewards of the Earth, but also because the quality of our lives depends on the environment.

Students analyze the interconnectedness among environmental systems and other major societal issues. They find direct connections and draw stock/flow diagrams and causal loop diagrams to show how progress in any of these connected systems can affect the others in positive ways.

Materials

- Reference articles, p 210
- Handout: Connections Among Interconnected Systems, p 179
- Resource: Intro to Stock/Flow Diagrams, p 223
- Resource: Intro to CLDs, p 220
- Resource: Intro to Connection Circles (optional), p 218
- Resource: Intro to the Iceberg (optional), p 226

Procedure

1. How are environmental issues part of a web of complex systems that affect each other and often make solutions to the world's problems appear so difficult? Share this diagram with

students. It shows four major categories of world systems with arrows connecting them (FIGURE 1).

FIGURE 1

2. How do these four systems affect each other? Distribute the handout. Assign teams to brainstorm how the six pairs of connections affect each other. Students may draw upon prior knowledge as well as doing more research. Be sure they emphasize the positive outcomes and potential progress made possible by the interlocking nature of systems.
   - Environmental systems ↔ Social justice
   - Economic prosperity ↔ Environmental systems
   - Public health ↔ Environmental systems
   - Public health ↔ Economic prosperity
   - Social justice ↔ Public health
   - Social justice ↔ Economic prosperity

Hope and opportunity amidst crisis

As the COVID-19 pandemic unfolded, a confluence of events affected the world with universal impact. While environmental crises continued, the pandemic spread a new virus across the globe, infecting hundreds of millions of people and killing millions. A huge segment of the international economy ground to a halt. Then the brutal police killing of George Floyd, an African-American man in Minneapolis, Minnesota, was recorded on video and seen around the world. More incidences of violence and injustice also came to light.

But crises can reveal pathways to change and raise awareness for the need to act. Protests and rallies in the US and other nations called for social justice and law enforcement reform after centuries of violence and repression toward people of color. With remarkable speed, scientists developed vaccines to protect people against COVID-19 and medical personnel bravely treated patients despite serious personal risk. Many governments and industries accelerated initiatives to address climate and pollution crises.

Educators occupy a unique position. They can help students realize that despite the difficult challenges facing us, they have many opportunities to shape a sustainable and prosperous future.

3. Students can use connection circles for brainstorming and then find and trace feedback loops to draw causal loop diagrams. The tables that follow show examples identifying causal connections. These are non-specific, general mental models to potentially flesh out as students do research and create systems diagrams.

Environmental systems ↔ Social justice

Positive effects	Negative effects
Green* initiatives improve quality of life for all.	Climate change impacts disadvantaged people more severely.
Renewable energy is cleaner and less expensive, helping to reduce poverty and inequality.	Polluting industrial sites are often located near housing for people powerless to resist.

Economic prosperity ↔ Environmental systems

Positive effects	Negative effects
Green jobs create employment and a higher standard of living.	Companies' need to make money for shareholders can drive pollution and short-term thinking.
Companies can use their resources for developing innovative, sustainable technologies.	Energy intensive and polluting industries add to greenhouse gas emissions and harm the environment.

Public health ↔ Environmental systems

Positive effects	Negative effects
A healthy environment is safer for all – wider access to clean air, water, and food.	People who work outdoors suffer effects of climate change on the job.
Open spaces improve quality of health, provide places to exercise and recharge, and reduce the warming effects of urban "heat islands."	Communicable diseases are more threatening to people in crowded environments with poor sanitation.

Public health ↔ Economic prosperity

Positive effects	Negative effects
Careers that promote public health are important to the economy.	Disadvantaged people often have more dangerous jobs.
Populations experiencing healthy living and working conditions are productive and innovative.	High levels of economic activity can increase volumes of waste and pollution.

Social justice ↔ Public health

Positive effects	Negative effects
People with opportunities for education and professional careers tend to be in better health.	Access to health care is unequal and can be prohibitively expensive.
Public agencies can provide services for all to improve overall public health.	People in poor health struggle to satisfy basic needs for themselves and their families.

Social justice ↔ Economic prosperity

Positive effects	Negative effects
Society benefits when more people have the chance to reach their potential.	Poor areas are often left out of infrastructure improvements such as public transit.
Businesses can "do the right thing" and add value to quality of life for customers, employees, and society in general.	People excluded from economic growth cannot use their talents productively and receive low wages.

* The word 'green' refers to categories of initiatives, jobs, and other activities that are beneficial, or at least less harmful, to the environment.

4. Tables like these reinforce the interconnected nature of complex systems. Improving environmental conditions creates jobs, raising overall economic prosperity and public health. Those gains move communities toward higher standards of social justice.

   Have students read the suggested articles or find other sources that describe the effects of environmental problems on other systems. Encourage them to build stock/flow diagrams that illustrate these connections. Students who are unfamiliar with stock/flow diagrams can use connection circles and causal loop diagrams to show relationships. Researching specific conditions will allow them to support their mental models with data and examples. The following sequence provides a general framework outlining how strengthening environmental systems through green jobs can improve economic prosperity, public health, and social justice.

Stock/Flow of Green Industries

These stock/flow diagrams (FIGURES 2 - 5) describe a mental model of how environmental systems and social systems are linked. While the graphic creates a useful diagram of one way of thinking, it lacks equations and consistent units that are necessary for creating a quantitative model. In order for a simulation to run and provide answers to "what if" questions, every element in the diagram needs to be defined mathematically. Stock/flow diagrams without those definitions function more like maps, guiding users to trace overall effects of rates and accumulations within systems.

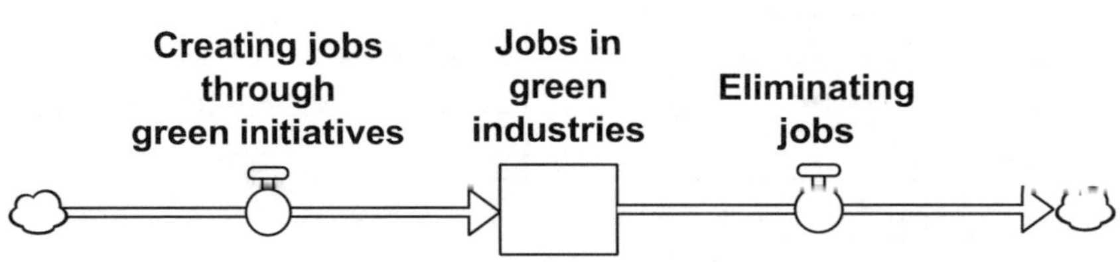

FIGURE 2

As environmental awareness grows, governments and industries will create jobs in green industries (FIGURE 2).

When more people are hired and work in green jobs, support for the initiatives that create those jobs will increase and drive new projects, powering a reinforcing loop. Money earned will increase also (FIGURE 3).

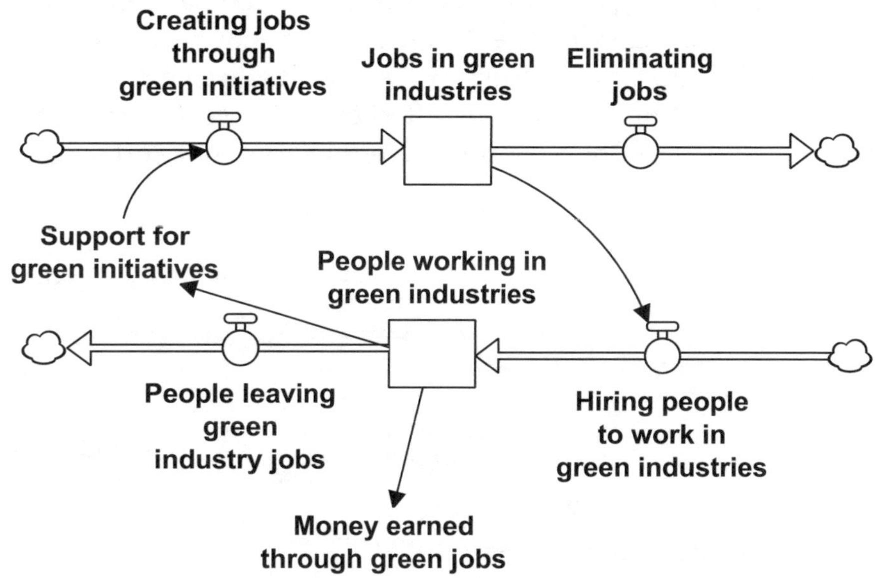

FIGURE 3

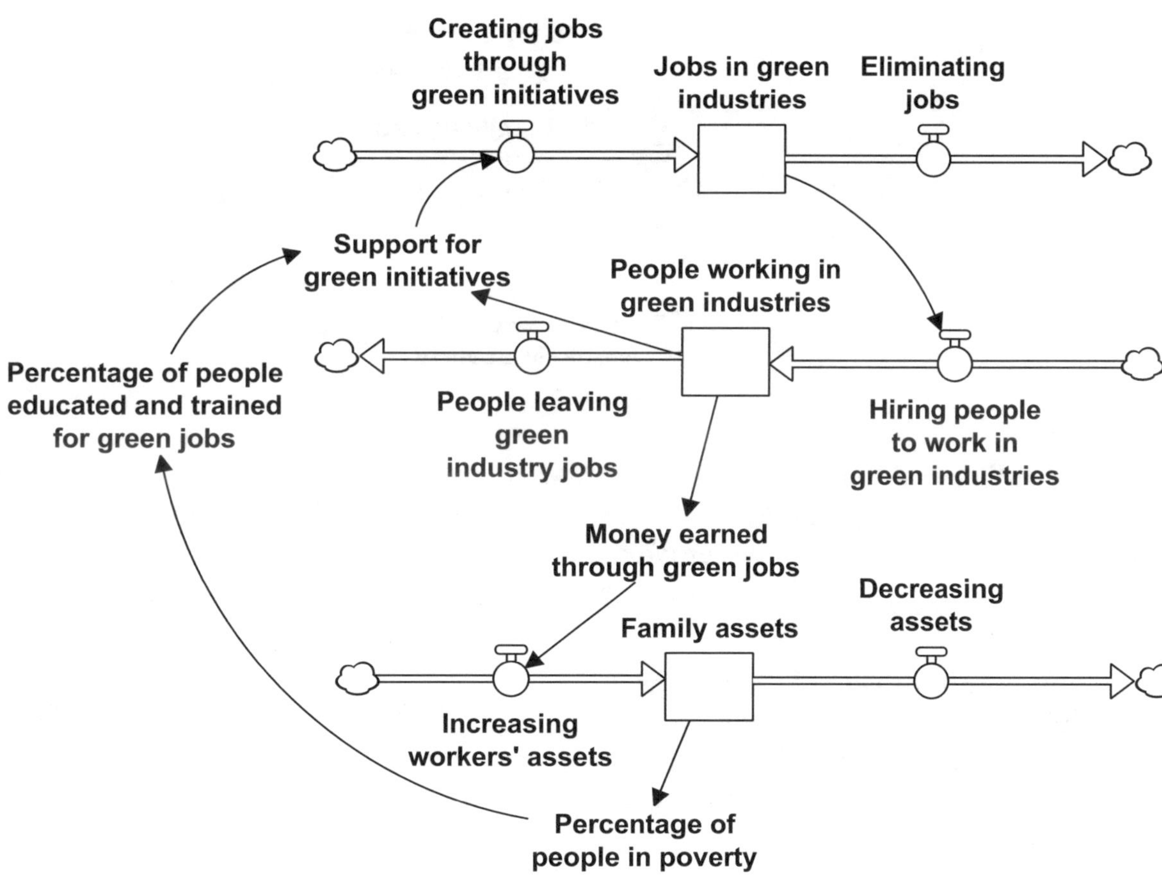

FIGURE 4

The money earned from green jobs will help pull families out of poverty. Family members will have more opportunities for education and training, increasing support for green initiatives. This reinforcing loop should benefit people from all backgrounds, improving conditions for racial justice and equality (FIGURE 4).

Families enduring conditions of poverty often have limited access to health care and nutrition, in addition to suffering the detrimental effects of environmental degradation where they live. Poor health can cause a drain on family assets due to medical expenses and inability to work and earn money. If families can be lifted out of poverty, public health will improve overall and create another feedback loop connecting environmental and social systems (FIGURE 5).

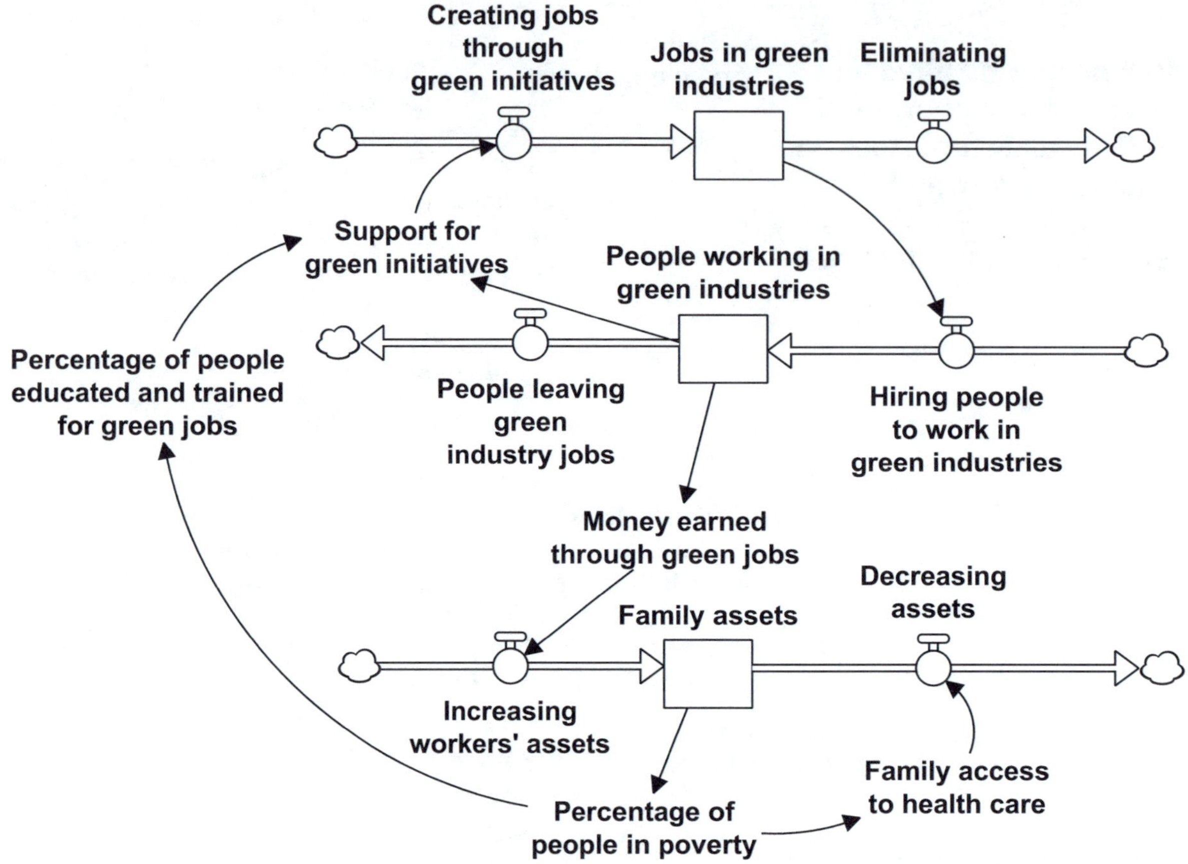

FIGURE 5

5. Students can pull causal loops directly from their stock/flow diagrams or choose to use different vocabulary and descriptions. The following examples show a sequence of possible feedback loops combining into a larger causal loop diagram. Like the stock/flow sequence in Step #4, the causal loop diagrams emphasize the potential for positive change when complex systems interact.

These causal loop diagrams (FIGURES 6, 7, and 8) contain no delay marks. Students will likely find roadblocks to optimistic ideas like these in their research. What can be done to move these improvements forward?

If "green" initiatives can mitigate effects of climate change, the people affected most severely will benefit economically. Overall economic growth will further increase support and funding for more green initiatives (FIGURE 6).

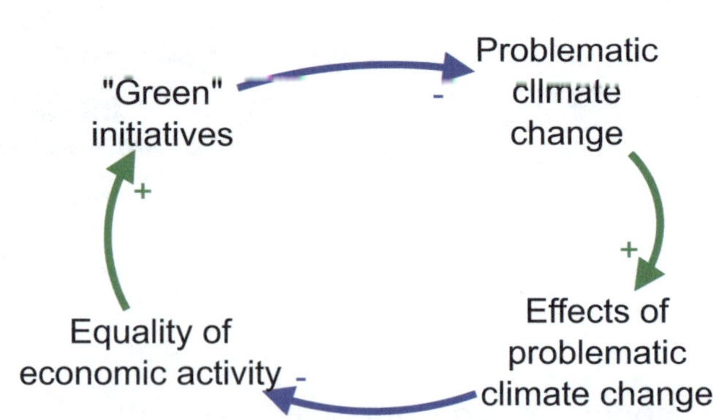

FIGURE 6

Generally, health will improve as the effects of climate change are reduced. Health care costs should decrease, freeing more money for green initiatives, and another reinforcing loop (FIGURE 7).

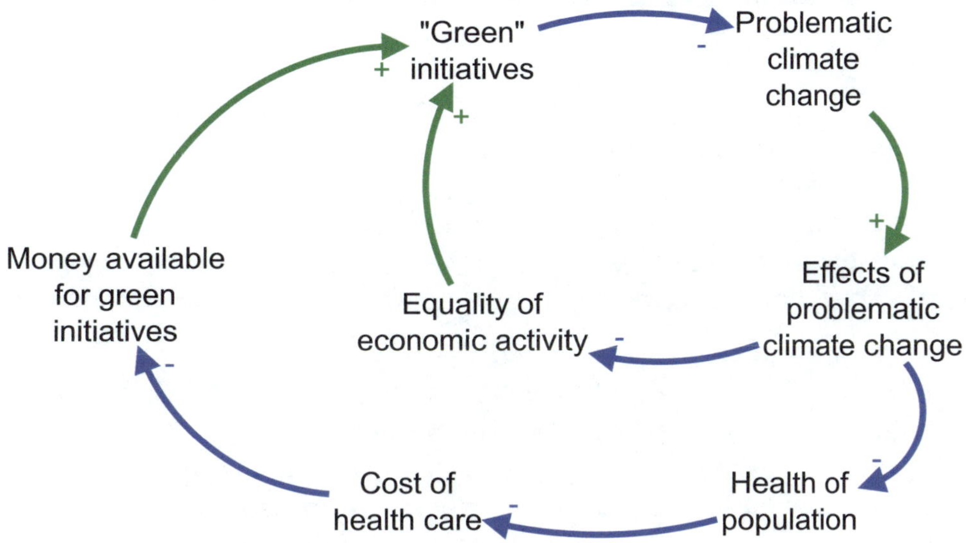

FIGURE 7

Green initiatives have the potential to drive additional reinforcing feedback loops (FIGURE 8).

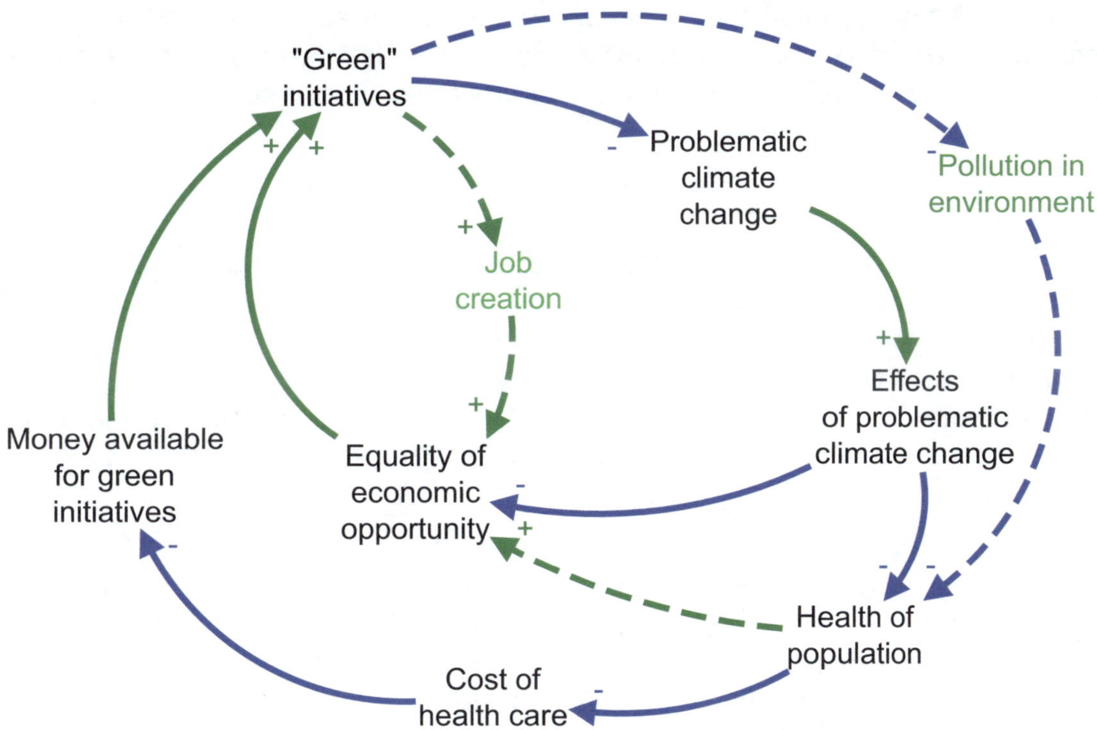

FIGURE 8

Green initiatives will create jobs, increasing economic opportunity, while reducing pollution in the environment. As overall health improves, more people are able to work and be more productive, growing economic opportunity.

6. Share causal loops and stock/flow diagrams on posters or projected as slides. Some students may expand the work begun in this activity for their independent project if you continue on to Lesson 5.2, Choosing a Topic.

Bringing the Lesson Home

How can improvement in any of the four systems (environmental systems, social justice, public health, economic activity) cause improvements in the others?

Students will have made many connections. A popular aphorism says, "A rising tide lifts all boats," and to some extent that applies to these complex systems. For example, a clean, sustainable environment should be good for everyone. But unless social justice increases, and fairness and equality of opportunity spread, many "boats" will be left grounded.

Many wilderness areas and animal and plant species are threatened. The rate of species extinction has accelerated in modern times. Can humans act to halt and reverse the decline of Earth's environmental systems?

Lesson 5.1 focuses on the interaction between environmental systems and social systems. Even without considering the effects on humans, taking care of the Earth remains a major goal of scientists, policy makers, and concerned citizens. Given time, nature can repair damage from natural and human-caused events. But once a species becomes extinct, it is lost forever.

One idea to preserve and restore ecosystems proposes to designate a percentage of Earth's area for non-human habitation and resource extraction. Another plan would reintroduce species to areas where their population has been lost or reduced. Students may choose to research environmental systems topics that focus on species protection in Lesson 5.2.

What hopeful signs are there that people can move complex systems toward improving people's quality of life?

Many serious problems exist, and the four systems considered in this activity are experiencing extreme pressure. However, progress has been made over time, and further improvements can power reinforcing feedback loops. Environmental awareness continues to grow, and people advocate for enforcement of laws protecting ecosystems and natural resources. Social justice champions work constantly to expand and safeguard individual rights around the world. Health care advances improve quality of life. Economic opportunities grow with increased availability of electricity, communications, and other technologies.

But the downside pressure on these systems remains intense. Reinforcing feedback loops can move in either of two directions, creating either virtuous or vicious cycles. An optimistic view sees incremental change over time, moving in the right direction, albeit with delays. Progress in complex systems is seldom linear and steady.

Everyone needs to maintain hope for a sustainable, prosperous future, especially young people. Educators must nurture the desire in students to protect and improve the world without understating the seriousness of its problems. Studying complex systems can help keep the feelings of urgency and possibility alive.

Handout: Connections Among Interconnected Systems

Look at the graphic and think about how these four systems affect each other. Use the table below to brainstorm how the six pairs of connections affect each other in both positive and negative ways. You can think about past experience or research your ideas. Students may draw upon prior knowledge as well as doing more research.

Try to find two positive and two negative connections for each link.

Environmental systems ↔ Social justice

Positive	Negative

Economic prosperity ↔ Environmental systems

Positive	Negative

Public health ↔ Environmental systems

Positive	Negative

Public health ↔ Economic prosperity

Positive	Negative

Social justice ↔ Public health

Positive	Negative

Social justice ↔ Economic prosperity

Positive	Negative

Which of your connections would you like to research further in order to create a diagram of that system?

5.2 Choosing a Topic
Independent Study of Environmental Systems

Environmental Systems
- What topic in environmental systems do you want to research?
- How does your topic affect your local community, and does it connect to other systems – social justice, public health, and economic prosperity?

Tools
- Student choice

Skills
- Varies, based on project and tools selected

"Think globally, act locally."
Patrick Geddes

Overview
Environmental systems affect everyone, and we all view them from a personal perspective. Weather and climate, wildlife, natural resources, food, ecosystems and habitats, recreation and leisure, … our daily lives involve multiple interactions with the natural world.

Encourage students to identify their most passionate interest involving environmental systems and to choose a topic for research on their own or with teammates. Students can explore local issues, creating an opportunity for them to engage with individuals and organizations in their communities. What policy decisions and strategies will be necessary to move toward long-term sustainability and a healthy environmental future, both locally and worldwide?

After students investigate their topics, they present reports using systems tools. Graphs, feedback diagrams, iceberg models, stock/flow diagrams, and other graphic representations can all contribute to explaining students' thinking. If computer modeling is part of your curriculum, students can build models of their findings and try out "what if" scenarios.

Materials
- Access to research materials (online, print, local organizations and companies)
- Systems Thinking/System Dynamics Project Standards, p 244
- Handout: Connections Among Interconnected Systems (optional from Lesson 5.1), p 179
- Handout: Systems Principles, p 212
- Handout: Summary of Systems Tools, p 213
- Handout: Project Planning Guide, p 189

Procedure
1. Review students' prior knowledge and their previous studies of environmental systems. This is a good time to summarize the common characteristics of environmental systems. If students have learned about Barry Richmond's Systems Thinking Skills, review those guideposts and how they facilitate understanding of complex problems.

2. Distribute the handout, "Project Planning Guide." As students choose topics, form teams with similar interests and have them investigate current situations in environmental systems. The handout from Lesson 5.1, "Connections Among Interconnected Systems," can be useful, helping students find topics to investigate more deeply.

   Independent and team study will likely fall into three overlapping categories:

   - New Topics - Specific issues they haven't studied yet
   - Going Deeper - Looking further into issues already raised in previous activities
   - Local Issues - Researching initiatives or topics in their own community or region

   Because topic choice can vary, teachers should be prepared to act as coaches, guiding teams as necessary. Reviewing systems thinking tools and principles will be useful for helping students navigate their way around the handout, "Project Planning Document." Short examples about employing "closed loop thinking" and "thinking backwards" are given on the following pages; those systems thinking basics are important components for studying complex situations.

3. Before students begin their projects, distribute the handout describing the systems tools they can use to organize and present their work. Here's a review of some of their options:

   - Behavior over time graphs – What variables are changing?
     Usually a good starting point – Find data where possible and be clear when graphs represent mental models versus research findings.
   - Connection circles – What elements in a system are affecting change in other elements directly?
     Good brainstorming tool for identifying important elements in a system and how they affect each other. Tracing causal connections can help find feedback loops.
   - Causal loop diagrams – Where are feedback loops in the system?
     Complex systems usually have "competing loops" – try to find reinforcing loops and balancing loops that may be in conflict. Can students find common ground for sustainable solutions?

Choosing Topics

Challenge students to identify issues they care about and that are important and necessary to address now. What are people doing locally, regionally, nationally, and globally? Here are some suggestions to stimulate student engagement:

- As a class, brainstorm a set of environmental issues. Give each student three colored dots to place next to the three topics they consider most important.
- Ask students to imagine the world 25 years in the future. They can write a short description or share their story in small groups. How did the Earth get that way? If their vision is a positive one, what needs to be done to achieve it? If they describe a negative picture, how can it be avoided?
- Have students create a short questionnaire about environmental systems issues. For homework, assign them to survey a family member or friend and share the survey results as a class. Questions may include the ones posed above about important environmental issues, what the future may look like, and what needs to be done.

- Archetypes – Can we identify a generic structure, a pattern or behavior that occurs in a variety of situations?
  Systems thinking helps us to transfer knowledge about a specific subject to other situations. Viewing behavior in a broader interdisciplinary context increases understanding and anticipates outcomes that may be counterproductive to one's goals.
- Iceberg diagrams – Looking beyond events, what are longer-term trends? What structures are driving those trends? What mental models and beliefs have contributed to building the iceberg ?
  An event is just "the tip of the iceberg." Complex systems problems resist short-term, linear thinking. To change behaviors, we need to change the structures driving them.
- Stock/flow diagrams – What is accumulating or diminishing in the system? What are the rates of change? Stocks are changed only by flows and their units must be the same. Stock/flow diagrams show how things actually work.
- Computer models and simulations – How can we test assumptions and try out "what if …" strategies?
  Model building demands rigor and precision, and the payoff is calculation beyond what could be done mentally as quickly and accurately.

Closed loop thinking differentiates systems thinking from linear thinking.

Encourage students to seek hopeful solutions to environmental systems problems by employing the important systems thinking skill of closed loop thinking.

Linear Thinking

Why do proposed solutions often fail to achieve the desired results? And why do problems return after a while, sometimes becoming worse or simply morphing into a set of different obstacles? Linear thinking (FIGURE 1) creates a chain leading to action taken to solve a problem.

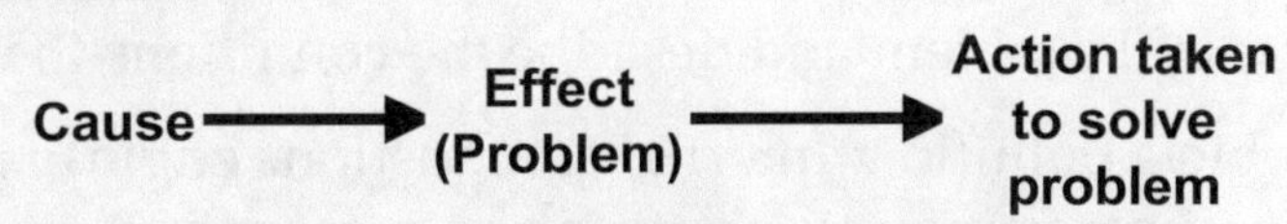

FIGURE 1

Increasing road capacity may seem like a reasonable way to reduce traffic congestion (FIGURE 2).

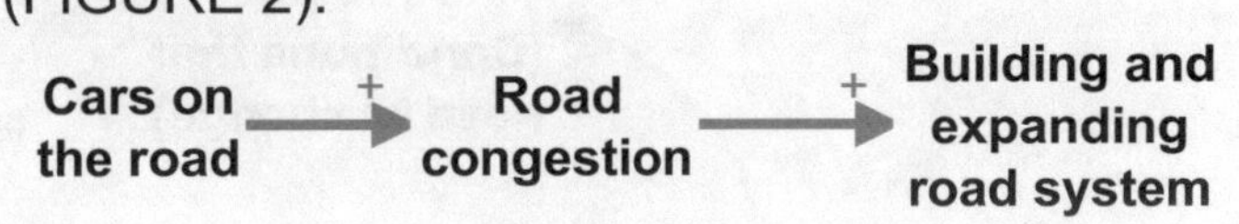

FIGURE 2

Closed-loop Thinking

Discovering feedback through closed loop thinking (FIGURES 3 - 4) can add insight to problems that allows for deeper understanding. The effect of the action taken attempting to solve the problem may make the problem worse:

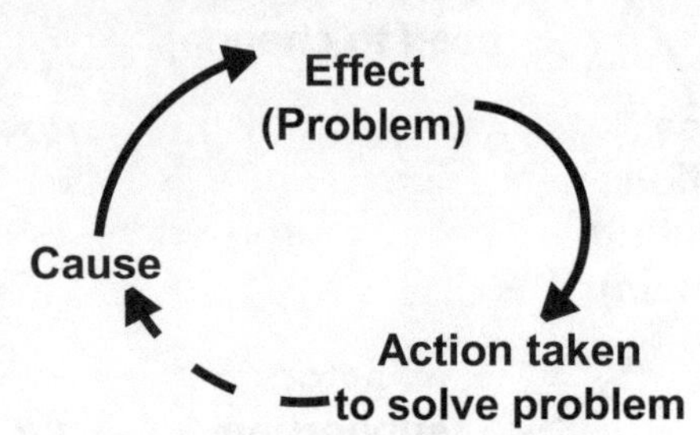

FIGURE 3

Closed loop thinking (continued)

When new or improved roads are developed, driving becomes a more attractive choice, and traffic will probably increase. Closed loop thinking helps us realize that the action taken will itself become a cause and affect the conditions that created the original cause. The goal may be to see the original cause fade away and the problem disappear, but often the opposite happens; the action taken exacerbates conditions.

Although maintaining a robust infrastructure for automobile traffic should be a factor, planners need to look beyond private vehicles commuting to solve road congestion problems. Public transportation, flexible work schedules, remote work options, and other alternatives can be more effective, better for the environment, and less disruptive than expensive, prolonged construction.

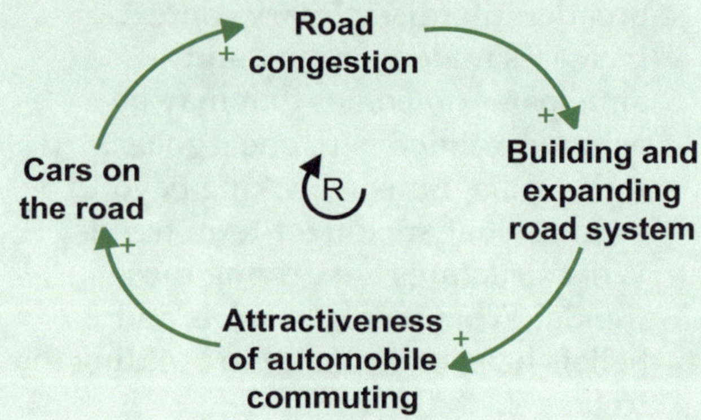

FIGURE 4

Thinking Backwards

Closing causal loops can be difficult. Here's where thinking backwards might help. Begin with the original problem and describe the conditions that create that problem, (FIGURE 5).

Vehicle pollution: Internal combustions engines powered by gasoline and diesel fuels emit significant amounts of greenhouse gases and other air pollutants. Thinking backwards can help create a causal loop diagram that describes how a change toward more electric vehicles ("Ratio of E-cars..." in FIGURE 6) might be achieved.

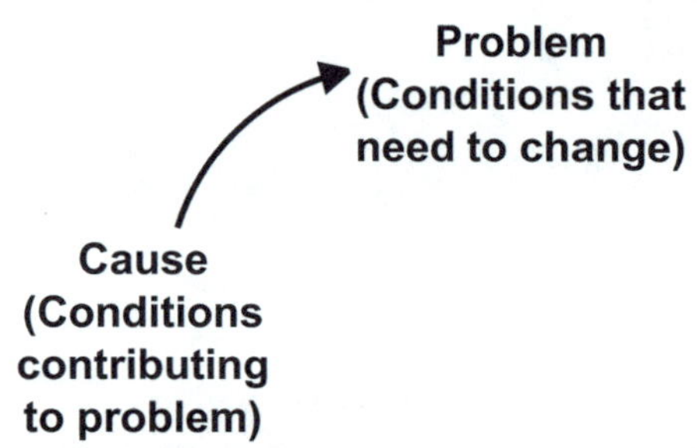

FIGURE 5

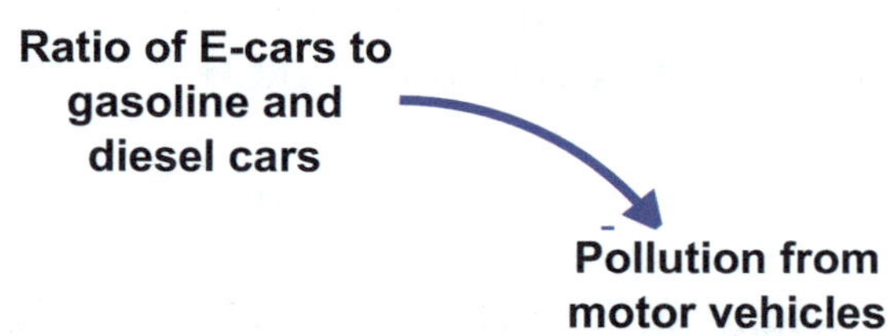

FIGURE 6

Until more motor vehicles operate without fossil fuels, pollution will be a problem. Then think backwards: what factors have allowed that cause to exist and become influential? (FIGURES 7 - 8)

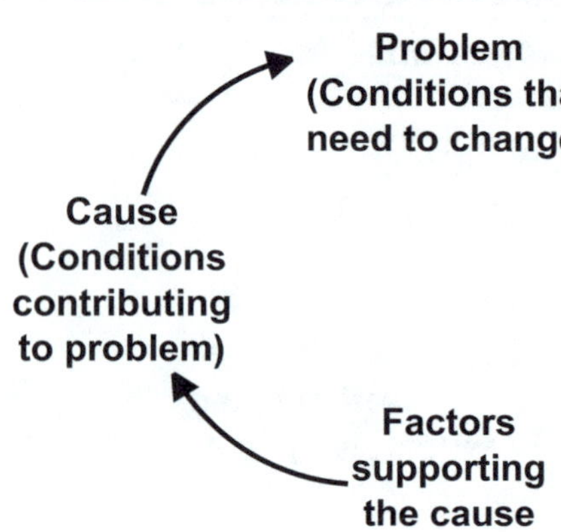

FIGURE 7

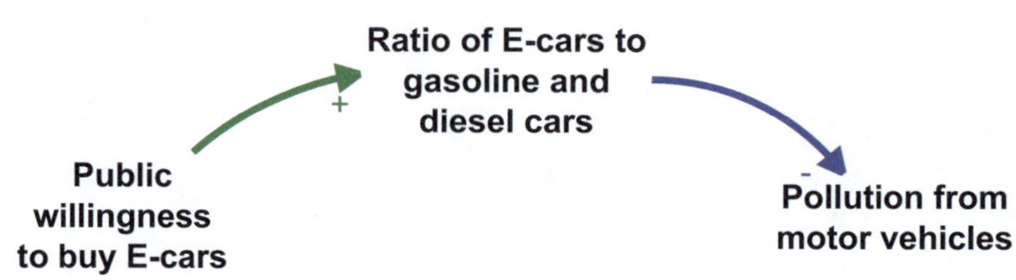

FIGURE 8

Historically, sales of electric vehicles remained low because of several factors. Raising people's willingness to buy E-cars will increase sales.

Now consider action steps to change those underlying factors (FIGURES 9 - 10).

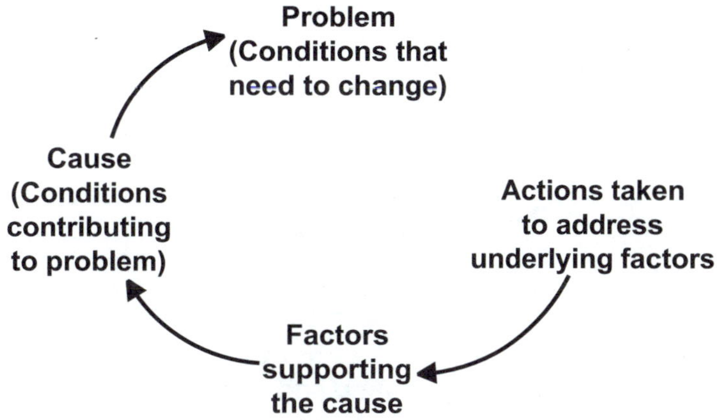

FIGURE 9

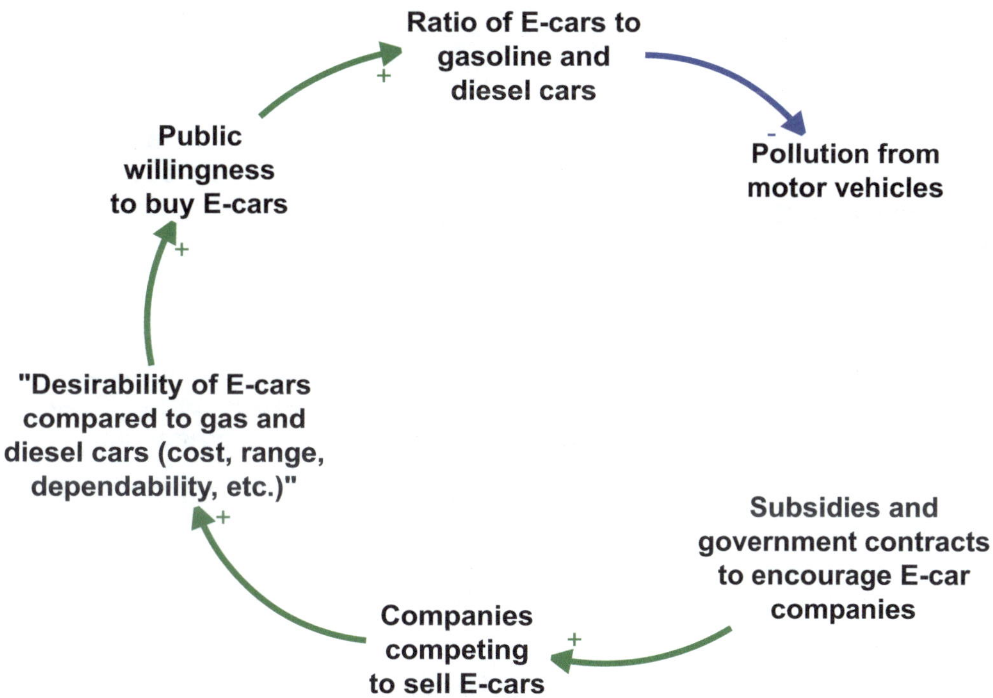

FIGURE 10

Subsidies for buyers of E-cars and other government actions increases the competition among companies building them, making ownership more desirable.

Because the problem created the need for actions, the loop can be closed (FIGURES 11 - 12) with one or two more connectors. The actions taken may not improve conditions enough to consider the problem "solved" and could cause unintended consequences, but thinking backwards has created a closed loop with a more realistic view of problem solving.

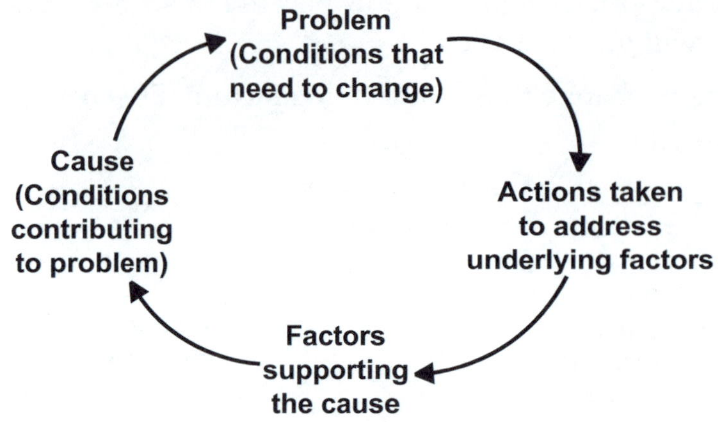

FIGURE 11

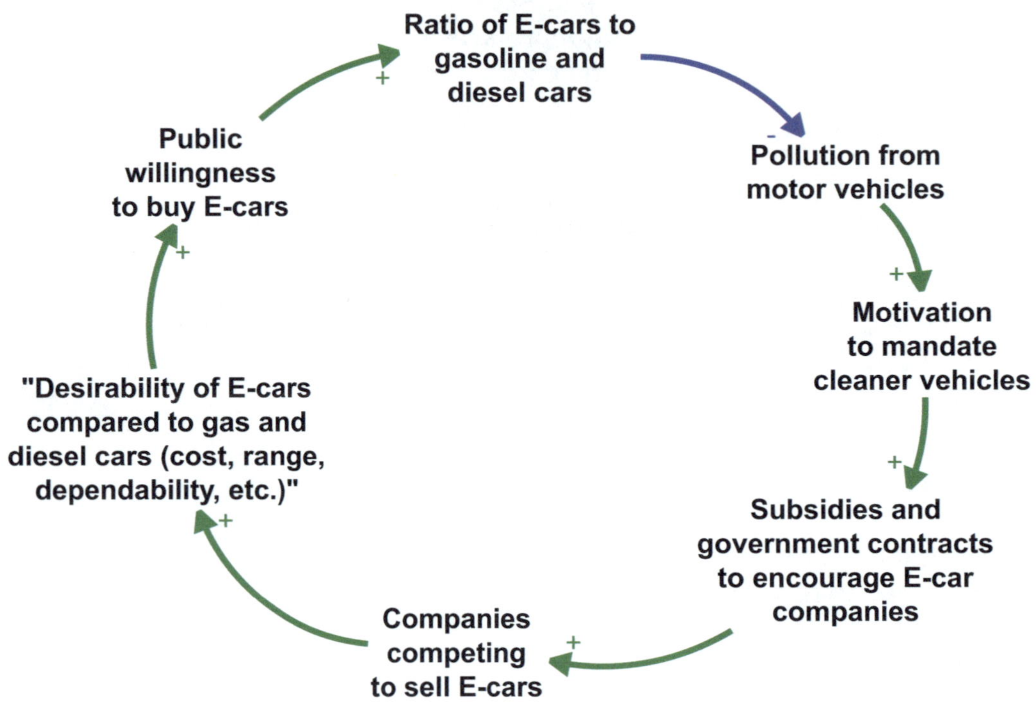

FIGURE 12

The problem of "Pollution from motor vehicles" created enough motivation for actions that expanded the number of E-cars.

When you have difficulty closing a loop, thinking backwards can help.

Bringing the Lesson Home

What is meant by the advice, "Think globally, act locally?"

Studying systems teaches us that many boundaries are arbitrary and permeable. The land, air, and water systems of the world are interconnected and affect social systems. All citizens of Planet Earth eat food from the land, breathe air, and drink water, so planning for a sustainable future requires a global perspective. Our actions often have the greatest effect in our own communities but eventually make a difference to the world as a whole. When we decide to act responsibly and make sure we protect the Earth, it benefits our local region as well as the planet as a whole.

Why did you study the problem you chose?

Students will have different reasons for choosing a problem to study. Share their stories about why a particular topic appealed to them. Maybe someone became concerned about receding glaciers during a family hike. Another person may value fishing with friends. Most students will have concerns about maintaining clean air and water. Everybody should have some connection with nature and appreciate the importance of understanding environmental systems.

In addition to the important goal of protecting and preserving natural systems, environmental issues reach deeply into other systems in society. Threats to environmental systems weaken social justice, public health, and economic prosperity. Students should be able to connect their topic to one or more other systems.

What do you see as your role as a citizen helping to shape environmental policy?

Change usually follows a two-step process. Awareness and understanding must come before action can be taken to address specific situations.

Awareness and understanding - After studying environmental systems and learning more about the history and complexity of problems, students will likely feel the need to educate themselves further as well as informing their fellow citizens. As they continue their own learning, they can find groups of people with similar goals and seek opportunities to share their models.

Action - "Think globally act locally." Students can join their peers in environmental clubs or groups. Many communities offer opportunities for them to join other citizens for action projects like trail maintenance and other acts of stewardship. Political action is possible through local government and elected officials. Student voices gain leverage as they approach and reach voting age.

How are your diagrams and models useful?

Visual tools are often effective in convincing people to act or support a cause. Video, photographs, factual diagrams, and charts can all have powerful effects on people. The systems tools students have used can all help to tell powerful stories.

Students will cite aspects of their diagrams and models that add insight to a situation. Perhaps they have identified leverage points, which Donella Meadows defined as "… places within a complex system … where a small shift in one thing can produce big changes in everything."

Bringing the Lesson Home
(continued)

Reversing one polarity (direction of change) in a causal loop diagram reverses the direction of all the other elements in the loop. Suggestions like replacing a certain number of cars on the road each day with enough buses to carry the commuters as passengers will slow down the release of CO_2 into the atmosphere.

Convincing drivers to forgo their own vehicles and ride in a bus may be difficult, but a diagram or model that shows the reduction in emissions can make a strong argument.

Dictionaries define literacy with two meanings; first, as the ability to read and write, and second, as knowledge or competency in some area. Citizens need environmental systems literacy under that second definition in order to become systems citizens. Barry Richmond coined that term to describe people who consider themselves members of a global community. They strive to understand the complexity of problems in the world and work to bring about positive change. Systems citizens participate in the decision process and policy-making necessary for a sustainable future. The path our society takes has important consequences, and each person has a right and responsibility to participate in the choices.

Did you draw a stock/flow diagram or create a working model? What aspects of building a model were helpful? What steps were difficult?

Translating a mental model to a stock/flow diagram pushes us to think about how things actually work. The units in the flows have to match the units in the stocks to which they are connected. When we try to get a model to run, our mistakes can be hard to correct. Besides the syntax of the software we're using, we have to make sure the logic in the model reflects our thinking. When we get a model running, we can change input values and ask "What if …" questions.

George Box wrote, "All models are wrong but some models are useful." What do you think that means and do you agree?

Any model is a representation of something and therefore isn't expected to be an exact copy. Models are designed to teach or predict something, not to duplicate reality. But the word "wrong" may be too strong to use when models are useful. A model can truly be labeled "wrong" if it is flawed to the point where it produces incorrect results or predictions. But accurate and useful models can educate and inform so that we can implement strategies to address even the most complex problems.

Handout: Project Planning Document

Note: You will need additional blank paper as you work on your project.

START HERE: ISSUE of concern (question or problem statement)

TRENDS
What's been happening? (graphs to show trends)

SYSTEM STRUCTURE
How does the system function? (connection circles, feedback loops, stocks and flows, and/or models to show key accumulations and interdependencies.)

LEVERAGE
How can you modify your diagrams to show ideas for improvement? (show below)

ACTION
If you took action, how did that go?

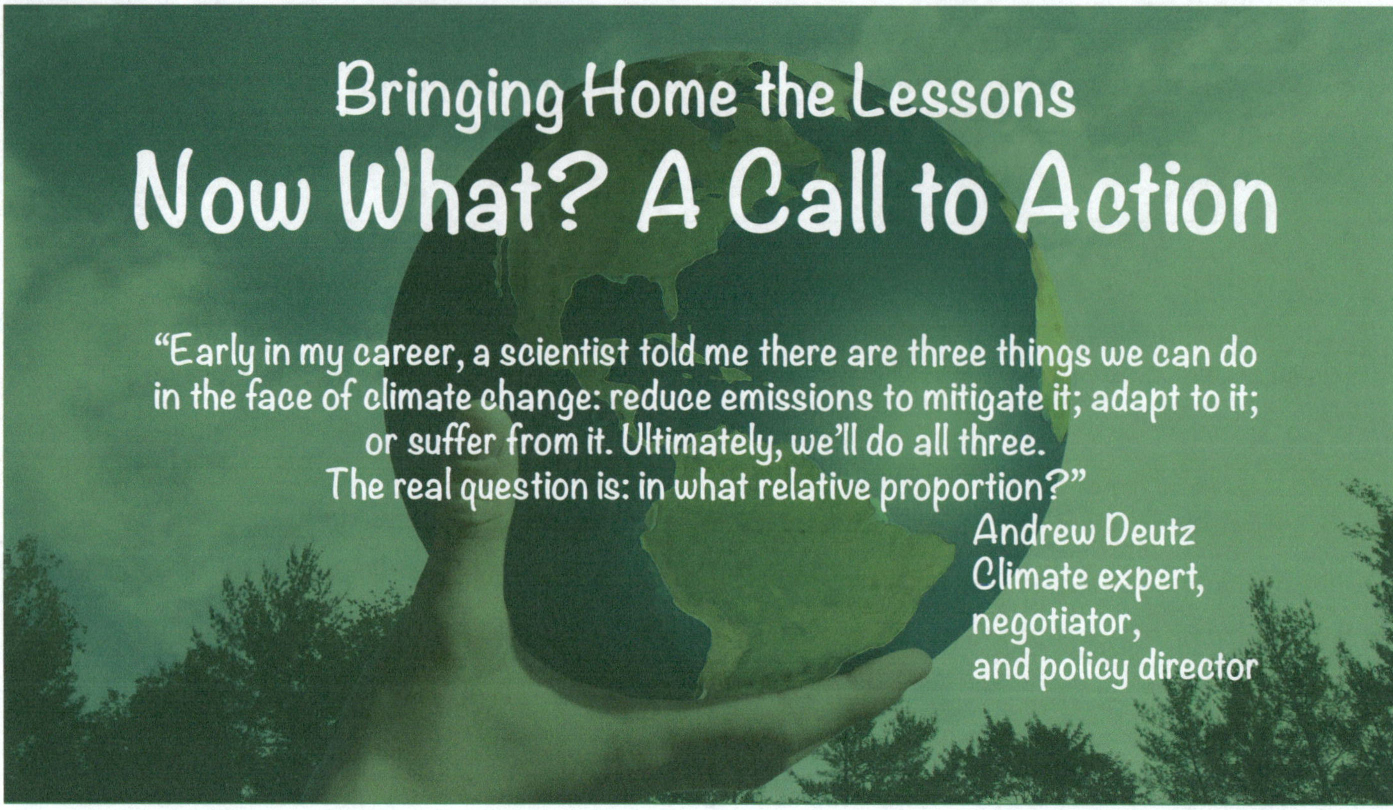

Bringing Home the Lessons

Now What? A Call to Action

"Early in my career, a scientist told me there are three things we can do in the face of climate change: reduce emissions to mitigate it; adapt to it; or suffer from it. Ultimately, we'll do all three. The real question is: in what relative proportion?"

Andrew Deutz
Climate expert,
negotiator,
and policy director

Students will be greatly affected by the answer to Dr. Deutz's question. Therefore, they must take a major role in determining how it gets answered. Lifestyle choices like recycling and using energy efficient lighting will make a difference. Even more importantly, global policies involving international agreements and major infrastructure initiatives will require the support of "systems citizens." Those who care about the effects of their actions and understand both long-term and short-term planning can become the leaders we need to make the 21st Century a time of hope and prosperity, justice and health, harmony and sustainability.

Having studied environmental systems, students can tackle the question, "Now what?" The answer will not be simple, and the path may not be smooth, but the time to begin is today.

Some categories for tackling "Now what?" as a systems citizen:

- Education – Keep on learning. Use your prior knowledge to find out more. Integrate what you can understand about science, social justice, economics, public health, and other fields.
- Information – Follow current events through reliable media.
- Local action – Improve the environmental systems in your community. Join groups that are working on issues you care about.
- Political action – Vote as soon as you are eligible. Support candidates who share your concerns.
- Lifestyle choices – Make decisions as a consumer and citizen that support sustainable policies wherever possible.
- Optimism – Enjoy the beauties and gifts of the environment. Your positive attitude will rub off on others and help them appreciate the Earth.

Appendix A: Connections

Overview

Many connections exist among the tools and lessons in the book and a multitude of educational standards at national, state, and local levels. Some national standards include:

- Next Generation Science Standards (NGSS) were created through the collaboration of multiple national organizations and 26 lead states in the US.
- Common Core State Standards were created through a state-led initiative in the US to establish a set of standards for English language arts and mathematics. Even though they are no longer in use in all states, they include important educational concepts. To see how they connect with systems education, visit the CLE website at http://www.clexchange.org/curriculum/standards/commoncore.asp
- STEM Education is an initiative to encourage a focus on science, technology, engineering, and math in K-12 education. To see how STEM learning goals connect with systems education, visit the CLE website at http://www.clexchange.org/curriculum/standards/stem.asp

Specific examples that follow show only connections between the lessons in the book and NGSS. Standards for mddle and high school are included, since they are the intended audience for the lessons. That said, teachers may find ways to adapt the lessons for younger audiences if desired. In addition, the content within the book is highly interconnected and included examples may apply across muliple chapters and lessons. As always, teachers will know best the standards that apply for their own situation.

Connections between Systems Thinking Tools and Crosscutting Concepts (NGSS)

Systems Tools	Crosscutting Concepts
	1. Patterns. 2. Cause and effect: Mechanism and explanation. 3. Scale, proportion, and quantity.
	2. Cause and effect: Mechanism and explanation. 3. Scale, proportion, and quantity. 5. Energy and matter: Flows, cycles, and conservation. 6. Structure and function. 7. Stability and change.
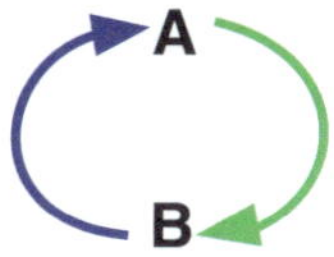	2. Cause and effect: Mechanism and explanation. 3. Scale, proportion, and quantity. 5. Energy and matter: Flows, cycles, and conservation. 6. Structure and function. 7. Stability and change.

Systems Tools	Crosscutting Concepts
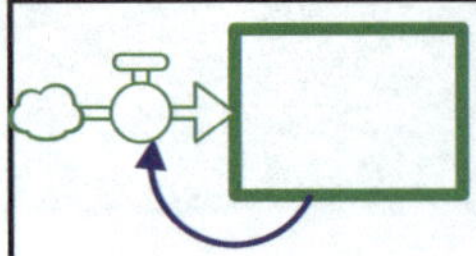	2. Cause and effect: Mechanism and explanation. 3. Scale, proportion, and quantity. 5. Energy and matter: Flows, cycles, and conservation. 6. Structure and function. 7. Stability and change.
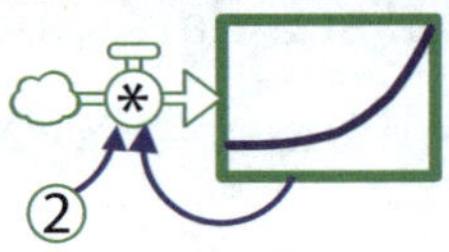	1. Patterns. 2. Cause and effect: Mechanism and explanation. 3. Scale, proportion, and quantity. 4. Systems and system models. 5. Energy and matter: Flows, cycles, and conservation. 6. Structure and function. 7. Stability and change.
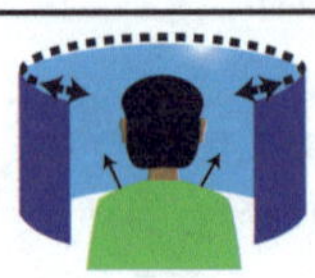	1. Patterns. 2. Cause and effect: Mechanism and explanation. 3. Scale, proportion, and quantity. 4. Systems and system models. 5. Energy and matter: Flows, cycles, and conservation. 6. Structure and function. 7. Stability and change.
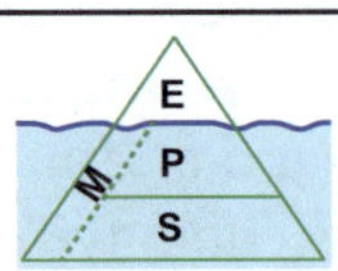	1. Patterns. 2. Cause and effect: Mechanism and explanation. 3. Scale, proportion, and quantity. 4. Systems and system models. 5. Energy and matter: Flows, cycles, and conservation. 6. Structure and function. 7. Stability and change.
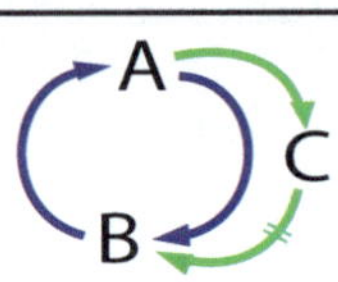	2. Cause and effect: Mechanism and explanation. 3. Scale, proportion, and quantity. 5. Energy and matter: Flows, cycles, and conservation. 6. Structure and function. 7. Stability and change.

Connections to the "Earth Space Science Progression" (NGSS)

The following tables show an overview of the book – chapters, lessons, systems tools, systems thinking skills, and example connections to NGSS.

	Lesson	Systems Tool(s)	Skill(s)	NGSS Connections
Chapter One: Sustainability, Systems Thinking, and the Environment	1.1 Sustainable Living in Cities			**ESS1.C The history of planet Earth** Grades 6-8 "Rock strata and the fossil record can be used as evidence to organize the relative occurrence of major historical events in Earth's history." Grades 9-12 "The rock record resulting from tectonic and other geoscience processes as well as objects from the solar system can provide evidence of Earth's early history and the relative ages of major geologic formations." **ESS3.C Human impacts on Earth systems** Grades 6-8 "Human activities have altered the biosphere, sometimes damaging it, although changes to environments can have different impacts for different living things. Activities and technologies can be engineered to reduce people's impacts on Earth." Grades 9-12 "Sustainability of human societies and the biodiversity that supports them requires responsible management of natural resources, including the development of technologies."
	1.2 Transportation			
	1.3 Electrical Energy	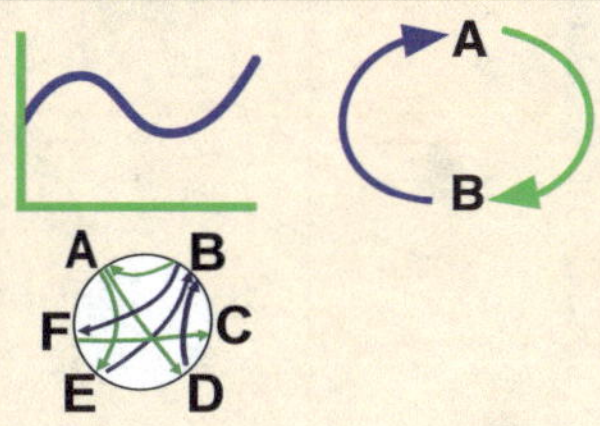	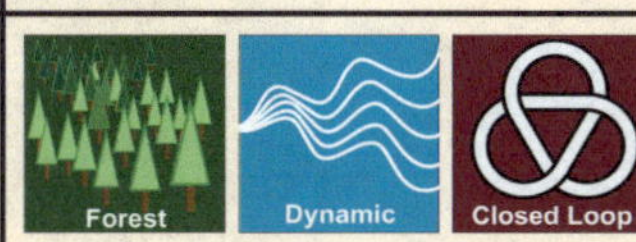	
	1.4 Recycling	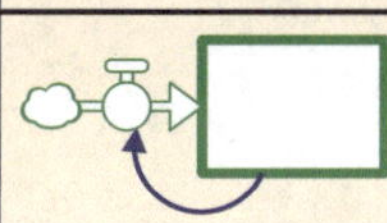		
	1.5 Environmental History			
	1.6 Modeling Population	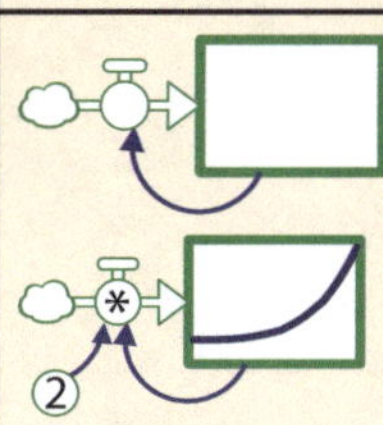		

	Lesson	Systems Tool(s)	Skill(s)	NGSS Connections
Chapter Two: Land	2.1 A Parable of Environmental Disaster	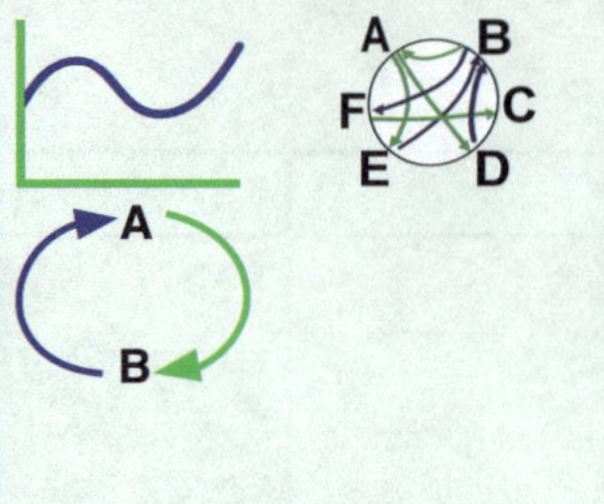		**ESS2.A Earth materials and systems** Grades 6-8 Energy flows and matter cycles within and among Earth's systems, including the sun and Earth's interior as primary energy sources. Plate tectonics is one result of these processes. Grades 9-12 Feedback effects exist within and among Earth's systems. **ESS3.A Natural resources** Grades 6-8 Humans depend on Earth's land, ocean, atmosphere, and biosphere for different resources, many of which are limited or not renewable. Resources are distributed unevenly around the planet as a result of past geologic processes. Grades 9-12 Resource availability has guided the development of human society and use of natural resources has associated costs, risks, and benefits.
	2.2 Tragedy of the Commons and Other Archetypes	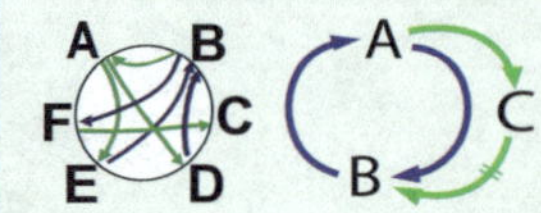		
	2.3 Protecting the Land During the Anthropocene Epoch	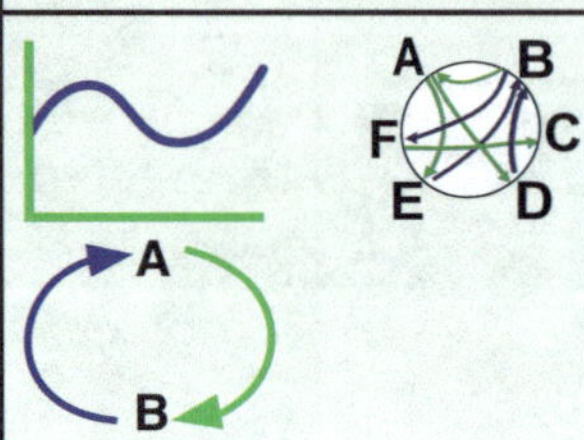		
	2.4 Cities and Urban Planning	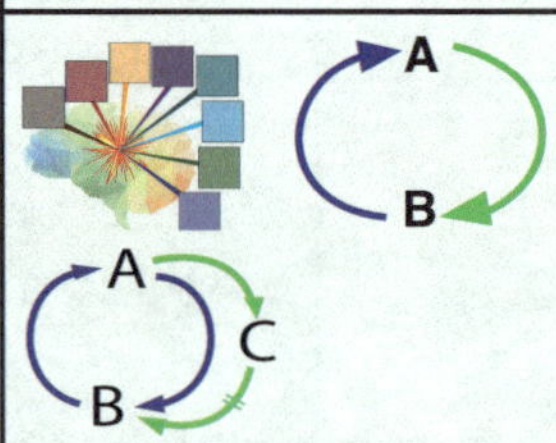		
	2.5 Studying Rapa Nui	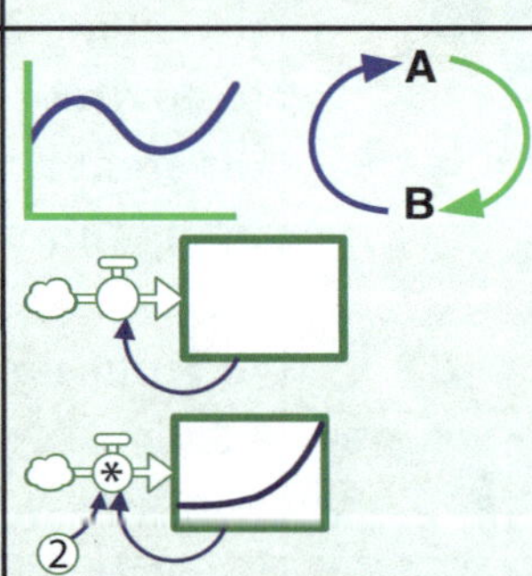		

	Lesson	Systems Tool(s)	Skill(s)	NGSS Connections
Chapter Three: Air	3.1 Protecting the Atmosphere	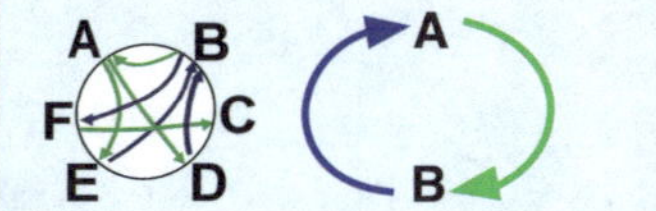	System-as-Cause, Closed Loop, Generic	**ESS2.D Weather and climate** Grades 6-8 Complex interactions determine local weather patterns and influence climate, including the role of the ocean. Grades 9-12 The role of radiation from the sun and its interactions with the atmosphere, ocean, and land are the foundation for the global climate system. Global climate models are used to predict future changes, including changes influenced by human behavior and natural factors. **ESS3.D Global climate change** Grades 6-8 "Human activities affect global warming. Decisions to reduce the impact of global warming depend on understanding climate science, engineering capabilities, and social dynamics." Grades 9-12 "Global climate models used to predict changes continue to be improved, although discoveries about the global climate system are ongoing and continually needed."
	3.2 Climate Change		Dynamic, Closed Loop, Empathic, Generic	
	3.3 Bathtub Thinking	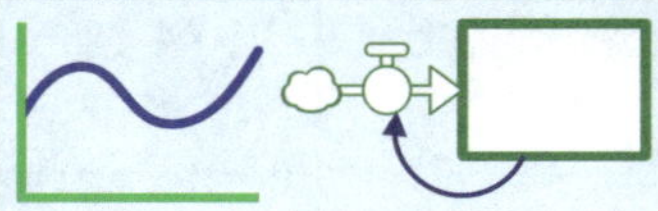	Dynamic, Operational, Scientific, Generic	
	3.4 The Carbon Cycle	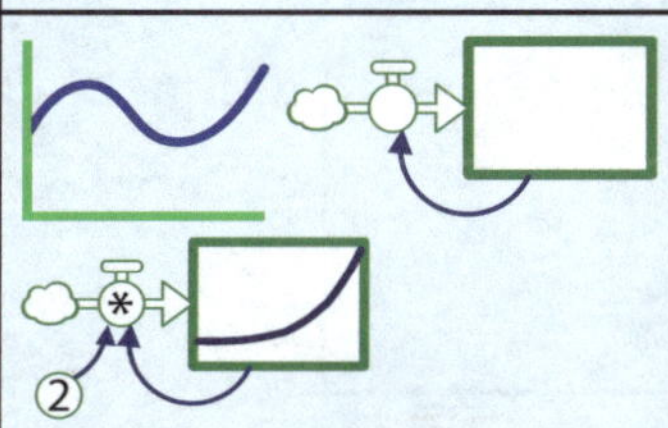	Forest, Dynamic, System-as-Cause, Operational, Empathic	
	3.5 Greenhouse Effect		System-as-Cause, Closed Loop, Scientific	
	3.6 Global Warming	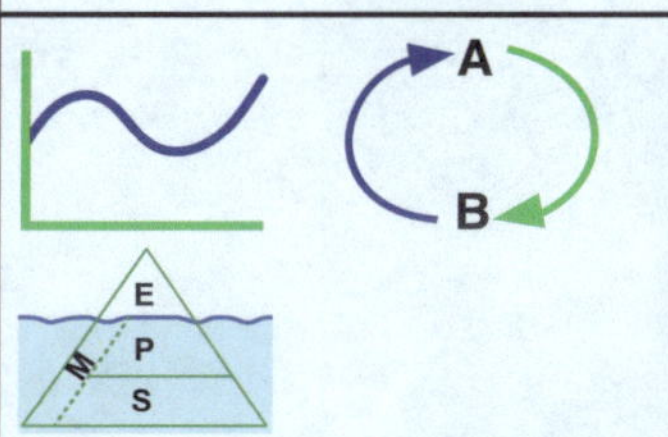	Dynamic, System-as-Cause, Closed Loop, Empathic	
	3.7 Examining Climate Models	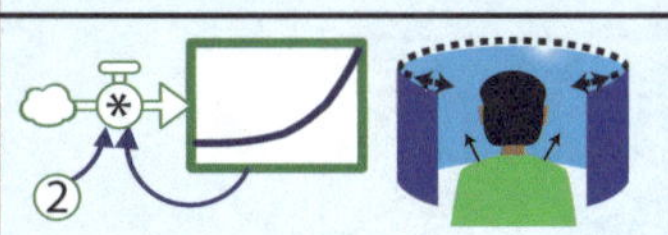	Dynamic, System-as-Cause, Scientific, Empathic	

	Lesson	Systems Tool(s)	Skill(s)	NGSS Connections
Chapter Four: Sea	4.1 Protecting Water	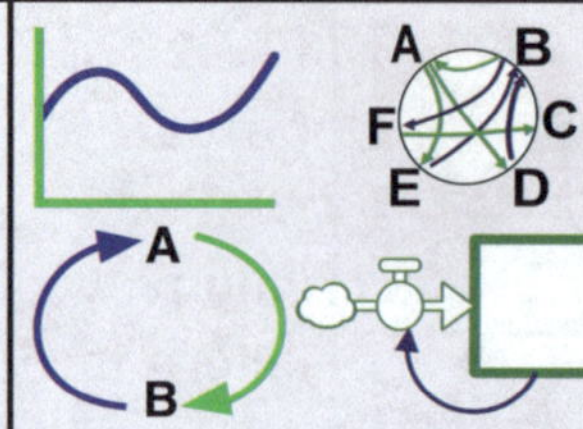	Dynamic, System-as-Cause, Operational, Closed Loop	ESS2.C **The roles of water in Earth's surface processes** Grades 6-8 Water cycles among land, ocean, and atmosphere, and is propelled by sunlight and gravity. Density variations of sea water drive interconnected ocean currents. Water movement causes weathering and erosion, changing landscape features. Grades 9-12 The planet's dynamics are greatly influenced by water's unique chemical and physical properties.
	4.2 Mosquito Nets and Overfishing		Closed Loop, Empathic, Generic	
	4.3 Sea Level Change	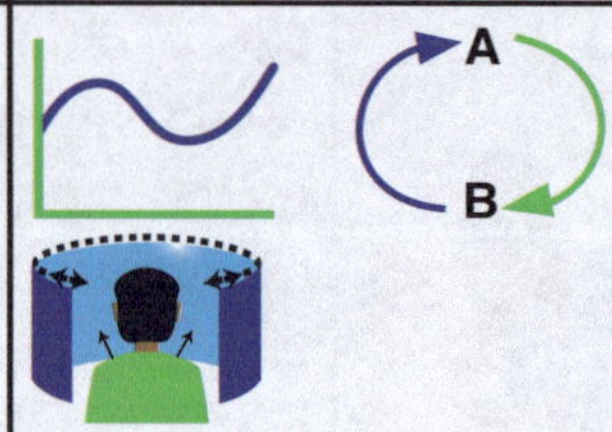	Dynamic, Closed Loop, Scientific	
	4.4 Altering the Ocean	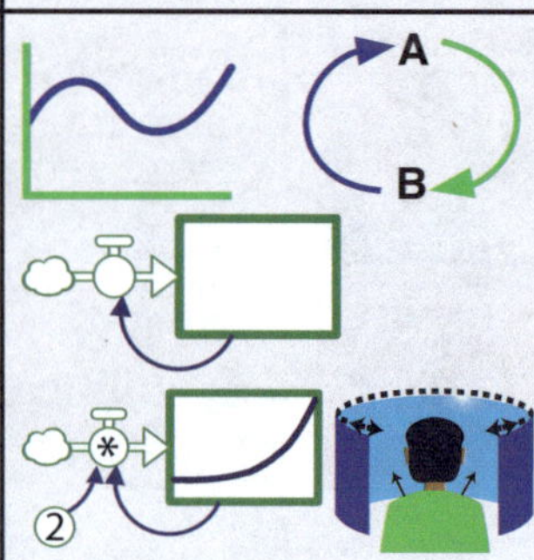	Forest, Dynamic, System-as-Cause, Closed Loop, Scientific	
	4.5 Running a Fishing Company	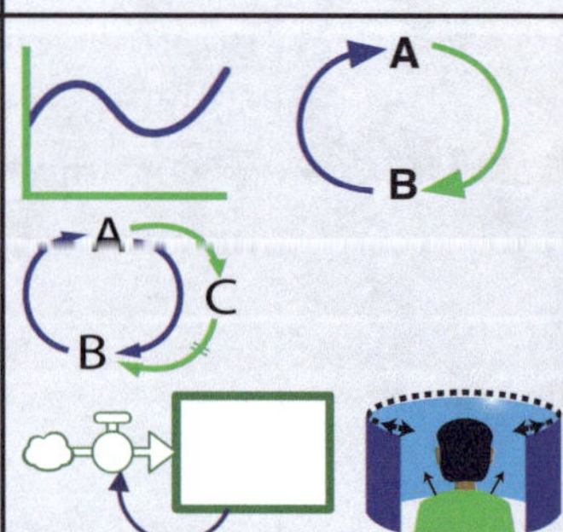	Forest, Dynamic, System-as-Cause, Operational, Closed Loop, Scientific, Empathic, Generic	

Appendix B: Lesson Resources

Overview

The materials listed within the lessons include links to suggested articles or websites. Keep in mind that articles and other online resources given as examples often change; links may stop working because they've been moved or deleted. You and your students may wish to explore and seek out additional sources of relevant information. Guide students to be mindful of bias within websites and urge them to use multiple sources, choosing credible sources that are known to check their facts.

The New York Times is cited frequently, as well as United States government websites and other sources. Many articles from commercial publishers are protected by a paywall or are available only to subscribers. After a certain number of articles read per month, non-subscribers aren't allowed to download and read any more. If you and your students are unable to access the suggested articles, here are suggested strategies for finding the information:

- Enter the article title or key words about the topic into a search engine. Articles may be reposted, and alternative reporting on the same topic should be available.
- Check school and local libraries for subscription privileges to periodicals.
- Investigate other media besides periodicals, like blogs, webinars, and video presentations
- Don't forget printed materials – books, magazines, newspapers, etc.

These resources also serve as a bibliography to illustrate where the author obtained information and ideas for the lesson plans, diagrams, and debriefing material.

Find the most recent links for all example data, articles, videos, and audio files here:
http://static.clexchange.org/curriculum/nowwhat/nowwhat_lessonresources_appendixB.pdf

Chapter One: Sustainability, Systems Thinking, and the Environment

Lesson 1.1 Sustainable Living in Cities

Articles about cities

- "Designing the City of Tomorrow"
  http://www.mcdonough.com/writings/designing-city-tomorrow-2017/
- "What Happened to the American Boomtown?"
  Growth patterns in American cities have changed as housing prices climbed.
  https://www.nytimes.com/2017/12/06/upshot/what-happened-to-the-american-boomtown.html
- "Sustainable Living"
  https://www.bbc.co.uk/bitesize/guides/zqvxdmn/revision/1
  Concise guide to attributes of sustainable communities.

Lesson 1.2 Transportation - Moving Efficiently

No additional examples

Lesson 1.3 Electrical Energy

Graphs of electricity use over time

- Energy data from The World Bank
  http://www.worldbank.org/en/topic/energy

General information about the difficulty of moving away from fossil fuels

- "The Long Slow Rise of Solar and Wind"
  Other writing and videos by and about Vaclav Smil provide summaries of his ideas concerning the slow pace of energy transformation.
  http://vaclavsmil.com/wp-content/uploads/scientificamerican0114-521.pdf

Articles about specific projects for electrical energy

- "The Power Plants That May Save a Park, and Aid a Country"
  Hydroelectricity brings economic development to the Democratic Republic of Congo, allowing conservation efforts to move forward.
  https://www.nytimes.com/2017/08/30/business/congo-power-plants-poaching.html
- "Hydropower for the habitat of mountain gorillas"
  https://www.climatepartner.com/en/carbon-offset-projects/hydropower-virunga-dr-congo
  Shows how renewable energy projects help protect the natural environment.
- "Elon Musk Just Built a Battery the Size of a Football Field in less than 100 Days"
  Short article contains video
  http://theweek.com/speedreads/740735/elon-musk-just-built-battery-size-football-field-less-than-100-days
- "Chile's Energy Transformation is Powered by Wind, Sun and Volcanoes"
  Includes video
  https://www.nytimes.com/2017/08/12/world/americas/chile-green-energy-geothermal.html
- "Good news from Chile"
  https://energytransition.org/2021/05/good-news-from-chile/
  Chile moves toward cleaner energy production.
- "The $3 Billion Plan to Turn Hoover Dam Into a Giant Battery"
  https://www.nytimes.com/interactive/2018/07/24/business/energy-environment/hoover-dam-renewable-energy.html
- "Hoover Dam as giant battery? The hurdles are more legal than technical."
  https://deeply.thenewhumanitarian.org/water/articles/2018/09/21/hoover-dam-as-giant-battery-the-hurdles-are-more-legal-than-technical
  Pumping water upstream may allow Hoover Dam to store energy.
- "The Biggest, Strangest 'Batteries'"
  Technologies for storing energy
  https://www.nytimes.com/2017/06/03/business/energy-environment/biggest-batteries.html
- "Electricity Storage"

https://www.epa.gov/energy/electricity-storage

- "How can we store renewable energy? 4 technologies that can help."
  https://www.weforum.org/agenda/2021/04/renewable-energy-storage-pumped-batteries-thermal-mechanical/

Lesson 1.4 Recycling

Articles, videos, or research materials about recycling

- Recycling in auto industry
  https://www.nytimes.com/2017/12/21/business/car-companies-recylcing.html
- "From grave to green: The fascinating world behind car recycling"
  https://gasp-pgh.org/2021/04/12/from-grave-to-green-the-fascinating-world-behind-car-recycling/
- Recycling basics
  https://www.epa.gov/recycle/recycling-basics
- Problems with plastic recycling [Opinion video]
  https://www.nytimes.com/2019/12/09/opinion/recycling-myths.html
- "Plastic Recycling is an Actual Scam" Video from "Climate Town"
  https://www.youtube.com/watch?v=PJnJ8mK3Q3g
- Examples of recycling ideas that work
  https://www.goodnewsnetwork.org/tag/recycling/

Lesson 1.5 Environmental History

No additional examples

Lesson 1.6 Modeling Population

Lesson plans from *The Shape of Change* - Note: "The Mammoth Game" requires dice

- "The In and Out Game" lesson plan
  http://static.clexchange.org/ftp/documents/x-curricular/CC2010-11Shape1InAndOutSF.pdf
- "The In and Out Game" simulation
  http://www.clexchange.org/curriculum/shapeofchange/soc_1_InOut.asp
- "Making Friends" lesson plan
  http://static.clexchange.org/ftp/documents/x-curricular/CC2010-11Shape2MakingFriendsSF.pdf
- "Making Friends" simulation
  http://www.clexchange.org/curriculum/shapeofchange/soc_2_Friendship.asp
- "The Mammoth Game" lesson plan
  http://static.clexchange.org/ftp/documents/x-curricular/CC2010-11Shape3MammothGameSF.pdf

Population simulations from Creative Learning Exchange

- "Rabbits, Rabbits, and More Rabbits" activity and simulation
  http://www.clexchange.org/curriculum/complexsystems/oscillation/Oscillation_PopulationC.asp
- "Why are there not More Elephants?" activity and simulation
  http://static.clexchange.org/ftp/documents/x-curricular/CC2012_Oscillations3BElephantsPopulationDynamics.pdf

Modeling simulation activity

- "Bumbles and Woofs"
  http://static.clexchange.org/ftp/documents/x-curricular/CC2018_BumblesWoofs.pdf

Chapter Two: Land

Lesson 2.1 A Parable of Environmental Disaster

Lesson plans

- "Lessons From the Lorax: Using Graphs to Study Change " and "Studying the Lorax with Feedback Loops"
  http://static.clexchange.org/ftp/documents/x-curricular/CC2012_TheLorax.pdf

Lesson 2.2 The Tragedy of the Commons

Explanations and examples of the Tragedy of the Commons archetype

- Primary source - "The Tragedy of the Commons" by Garrett Hardin – long, difficult, and technical
  http://static.clexchange.org/ftp/documents/x-curricular/CC1968_TragedyofCommons.pdf
- Find summaries, videos, etc. at student level via internet search.
  Example of video summary:
  https://ed.ted.com/lessons/what-is-the-tragedy-of-the-commons-nicholas-amendolare#review

Information about Bears Ears National Monument

- Fact sheet for Bears Ears
  https://www.fs.fed.us/sites/default/files/bear-ears-fact-sheet.pdf
- Department of the Interior interim report about reducing the size of Bears Ears Monument
  https://www.scribd.com/document/351066813/Interim-Report-EO-13792
  Summary at the beginning can be enough information for this activity
- News reports about reduction plan for Bears Ears
  Report 1: https://www.cbsnews.com/news/trump-bears-ears-national-monument-reduction-plan-utah/
  Report 2: https://news.nationalgeographic.com/2017/12/trump-shrinks-bears-ears-grand-staircase-escalante-national-monuments/
  Report 3: https://www.pbs.org/newshour/politics/biden-to-review-trumps-changes-to-national-monuments
  Report 4: https://apnews.com/article/donald-trump-joe-biden-utah-environment-and-nature-b763897fc47136339bd12360a90079c0
  Report 5: https://www.washingtonpost.com/climate-environment/2021/10/07/bears-ears-biden-monument/

Suggestion for reading at end of activity

- Donella Meadows "The Tragedy of the Commons on Georges Bank, and Elsewhere"
  http://www.donellameadows.org/archives/the-tragedy-of-the-commons-on-georges-bank-and-elsewhere/
  Spoiler alert – the author describes the "Fishbanks" simulation, so if you plan on playing that in class, only distribute the first part of the article or save to read after playing Fishbanks,"Activity 4.5.

Lesson 2.3 Protecting the Land During the Anthropocene Epoch

Geologic timelines and information about the Anthropocene Epoch

- "When Did the Anthropocene Begin?" Scientific American September 2016 p.36
  https://projects.katiepeek.com/Welcome-to-the-Anthropocene

Behavior over time graphs showing the human effects on the Earth, marking a new geologic epoch.

- "Anthropocene"
  https://www.nationalgeographic.org/encyclopedia/anthropocene/
- "Welcome to the Anthropocene" website with video
  http://www.anthropocene.info/

Articles about environmental threats to land ecosystems

Five Examples:

- "A Vigilant Rwanda is Winning the War on Plastic Bags"
  https://www.seattletimes.com/nation-world/a-vigilant-rwanda-is-winning-the-war-on-plastic-bags/
  African nation outlaws plastic bags, imposing strict penalties.
- "Brazil Wavers on Environment, and Wetlands Start to Wither"
  https://www.nytimes.com/2017/12/23/world/americas/brazil-pantanal-wetlands-michel-temer.html
  Development encroaches into forested land.
- "3 ways Brazil's environmental decisions affect the world"
  https://www.conservation.org/blog/3-ways-brazils-environmental-decisions-impact-the-world

Brief description of environmental challenges in Brazil

- "Monsanto's Weed Killer, Dicamba, Divides Farmers" New York Times 9/22/2017
  https://www.nytimes.com/2017/09/21/business/monsanto-dicamba-weed-killer.html
  Herbicide spreads onto neighboring fields, harming non-resistant crops.
- "Pesticide police, overwhelmed by Dicamba complaints, ask EPA for help"
  https://www.npr.org/sections/thesalt/2020/02/06/800397488/pesticide-police-overwhelmed-by-dicamba-complaints-ask-epa-for-help
  Pesticide sprayed on fields may spread to other locations. Includes link to audio
- "Vikings Razed the Forests. Can Iceland Regrow Them?"
  https://www.nytimes.com/interactive/2017/10/20/climate/iceland-trees-reforestation.html
  Efforts to restore areas that were deforested over one thousand years ago.
- "Iceland tries to bring back trees razed by the Vikings"
  https://phys.org/news/2019-07-iceland-trees-razed-vikings.html
- "The Birds and the Bees"
  https://foe.org/birds-bees-new-research-implicates-pollinator-toxic-pesticides-hummingbird-declines/
  Pesticides imperil hummingbirds and other pollinators.

Lesson 2.4 Cities and Urban Planning

Articles and videos about generic structures of cities compared to other systems

- "A Physicist Solves the City" full article
  https://www.nytimes.com/2010/12/19/magazine/19Urban_West-t.html
  Summary of "A Physicist Solves the City"
- Video – "The Surprising Math of Cities and Corporations"
  https://www.ted.com/talks/geoffrey_west_the_surprising_math_of_cities_and_corporations
- Shorter video – "Geoffrey West Explains Universal Scaling"
  https://www.youtube.com/watch?v=EXpOOg9fMHY
- Speech by Jay Forrester about generic structures in complex systems
  http://static.clexchange.org/ftp/documents/roadmaps/RM8/D-4434-1.pdf
  pp. 14 - 18

Lesson 2.5 Studying Rapa Nui

Videos

- "The Fate of Easter Island"
  http://www.pbs.org/wgbh/nova/earth/fate-of-easter-island.html

Articles about the history of Rapa Nui

- "The Lessons of Easter Island" by Clive Ponting
  http://blackville.nbed.nb.ca/sites/blackville.nbed.nb.ca/files/noteattach/y2016/Mar/mar._3_the_lessons_of_easter_island_reading.pdf
- "What Really Happened to the People of Easter Island"
  https://www.sbs.com.au/guide/article/2018/02/06/what-really-happened-people-easter-island
- "What Happened on Easter Island – A New (Even Scarier) Scenario"
  http://www.npr.org/sections/krulwich/2013/12/09/249728994/what-happened-on-easter-island-a-new-even-scarier-scenario
- "The Fate of Easter Island"
  http://www.pbs.org/wgbh/nova/earth/fate-of-easter-island.html

Articles about modern times on Rapa Nui

- "Rapa Nui is Remote. And Then There's Its Isolated North Coast"
  https://www.nytimes.com/2016/11/13/travel/rapa-nui-easter-island-north-coast.html
- "Rapa Nui and other fragile lands rally against tourism pollution"
  https://www.pbs.org/independentlens/blog/rapa-nui-and-other-fragile-lands-rally-against-tourism-pollution/
- "Easter Island is Eroding"

https://www.nytimes.com/interactive/2018/03/14/climate/easter-island-erosion.html

- "Easter Island limiting tourism as erosion puts its famous statues at risk"
  https://www.thetravel.com/easter-island-limiting-tourism/
  Challenges for Rapa Nui and other lands concerning development

Modeling lesson plans

- "Easter Island Population Model" Lesson plan from Diana Fisher
  http://static.clexchange.org/ftp/documents/social-studies/SS1994-09EasterIslandPop.pdf
- *Model Mysteries* Chapter 5 "The Disappeared" Lesson plan by Anne LaVigne and Lees Stuntz
  http://static.clexchange.org/ftp/documents/modelmysteries/Model%20Mysteries%20Book.pdf

Chapter Three: Air

Lesson 3.1 Protecting the Atmosphere

Background information

- Video - "How Does Air Pollution Affect Health?"
  https://blogs.scientificamerican.com/plugged-in/video-how-does-air-pollution-affect-health/
- Map of pollution sources - "The Top-22 Air Polluters Revealed"
  https://www.scientificamerican.com/article/the-top-22-air-polluters-revealed/

Articles about air pollution

- "Smothered by Smog, Polish Cities Rank Among Europe's Dirtiest"
  https://www.nytimes.com/2018/04/22/world/europe/poland-pollution.html
  Burning coal causes air pollution in Poland. The government has implemented programs to modernize heating systems.
- "Poland launches "clean air" program"
  https://www.trade.gov/market-intelligence/poland-launches-clean-air-program
- "Ship Exhaust Makes Oceanic Thunderstorms More Intense"
  https://news.agu.org/press-release/ship-exhaust-makes-oceanic-thunderstorms-more-intense/
- "New Delhi, Its Air Toxic, Declares a Health Emergency and Closes Schools"
  https://www.nytimes.com/2019/11/01/world/asia/delhi-pollution-health-emergency.html
  Dangerous air pollution creates a serious health crisis in India's capital city.
- "Delhi, the world's most air polluted capital, fights back"
  https://www.brookings.edu/blog/future-development/2020/11/25/delhi-the-worlds-most-air-polluted-capital-fights-back/
- "Popular National Parks Experience as Much Ozone Pollution as Major Cities"
  https://www.pbs.org/newshour/nation/popular-national-parks-experience-as-much-ozone-pollution-as-major-cities
  or https://www.scientificamerican.com/article/visiting-a-national-park-this-summer-hold-your-breath/
- Connection to plastic pollution (not diagrammed in lesson plan)
  "Where's Airborne Plastic? Everywhere, Scientists Find."
  https://www.nytimes.com/2020/06/11/climate/airborne-plastic-pollution.html
  Scientists have discovered airborne microplastics exist even in remote parts of the world.
- Opinion "You're Probably Inhaling Microplastics Right Now"
  https://www.nytimes.com/2020/06/25/opinion/plastic-air-pollution.html
  Environmental scientist warns that we must reduce the amount of plastic, which has become ubiquitous in modern civilization.
- "Thousands of Tons of Microplastics are Falling from the Sky"
  https://www.scientificamerican.com/article/thousands-of-tons-of-microplastics-are-falling-from-the-sky/

3.2 Climate Change

Background information about climate change

- Frontline video "Climate of Doubt"
  https://www.pbs.org/wgbh/frontline/film/climate-of-doubt/

Documentary explores how organizations spread doubt about climate change.

- Video: "How Global Warming Stacks Up"
  https://climate.nasa.gov/climate_resources/144/video-how-global-warming-stacks-up/
- Home page of climate and energy resources
  https://cleanet.org/index.html
- Video "Climate Change Basics"
  https://www.youtube.com/watch?v–ScX29WBJI3w#at=81
- Set of introductory lessons
  http://www.explainingclimatechange.ca

Audio and links to additional articles

- "Time Is Running Out To Control Climate Change, U.N. Report Says"
  http://www.wbur.org/onpoint/2018/10/09/climate-change-un
- "Here's How Far the World Is From Meeting Its Climate Goals"
  https://www.nytimes.com/interactive/2017/11/06/climate/world-emissions-goals-far-off-course.html
  Graphs and analysis of CO2 emissions around the world from 1990 projected to 2030.
- "Climate Change is Complex. We've Got Answers to Your Questions."
  https://www.nytimes.com/interactive/2017/climate/what-is-climate-change.html
  Overview of climate change facts presented in a question-and-answer format.

Articles about wildfires

- "Fire on the Mountain: 2 Forests Offer Clues to Yellowstone's Fate in a Warming World"
  https://www.nytimes.com/2017/09/13/climate/yellowstone-western-fires-in-two-forests.html
  Comparing two areas of forest that burned in Yellowstone National Park in 2000
- "Resilience of Yellowstone's forests tested by unprecedented fire"
  https://news.wisc.edu/resilience-of-yellowstones-forests-tested-by-unprecedented-fire/
  Scientists study changing conditions involving forest fires.
- Book review podcast "The Fire Next Time"
  https://www.nytimes.com/2017/12/29/books/review/the-fire-next-time.html
  Text of book reviews discussed in podcast.
- "Are American West's Wildfires Inevitable?"
  https://www.nytimes.com/2017/12/12/books/review/megafire-michael-kodas-american-wildfires.html
  Review of two books about the increasing size and threats of wildfires.
- "Why the American west's 'wildfire season' is a thing of the past - visualized"
  https://www.theguardian.com/world/2021/oct/10/wildfire-climate-emergency-us-west
  Climate change is suspected as a cause of increasing fire occurrence.
- "An Uncertain Future"
  https://www.npca.org/articles/1622-an-uncertain-future
  How climate change is affecting Mesa Verde National Park.
- "Australia's Wildfires Have Now Been Linked to Climate Change"
  https://www.sciencenews.org/article/australia-wildfires-climate-change

Lesson 3.3 - Bathtub Thinking

Materials to introduce stocks and flows

- Lessons 1 and 2 from *The Shape of Change* are helpful for beginners and for review.
  http://www.clexchange.org/cleproducts/shapeofchange_lessons.asp
- These worksheets from starting lessons in system dynamics (p. 11, p.16) help students distinguish between stocks and flows
  http://static.clexchange.org/ftp/documents/system-dynamics/SD1996-03GetStarted5Lessons.pdf
- Large problem set of stock/flow exercises with answer sheet attached. Look through the set and choose any pages for students that you think are useful:
  http://static.clexchange.org/ftp/documents/x-curricular/CC2013_

EverydayStocksandFlowsProblemSets.pdf

The Department Store Task

- Academic paper. Third page contains the problem. Do not reveal the answer, which is in the text on that page. Only reproduce the graph and questions for distribution to students: http://jsterman.scripts.mit.edu/docs/Cronin-2009-WhyDontWellEducated.pdf
- Department Store Problem is on p.48 of this paper: "Why don't well-educated adults understand accumulation? A challenge to researchers, educators, and citizens." http://web.mit.edu/jsterman/www/CroninGonzalezSterman061210.pdf
- Blog post including problem, hints, and answers. Be sure not to reveal solution until students have given their answers: http://www.generationaldynamics.com/pg/ww2010.i.macro061025.sterman.htm

Diagramming atmospheric CO_2 as a stock/flow diagram

- Bathtub dynamics – Atmospheric stock/flow diagram of CO_2 https://climateinteractive.files.wordpress.com/2009/11/ngbathtub.jpg

3.4 The Carbon Cycle

Carbon cycle and greenhouse gas information

- Atmospheric carbon dioxide https://www.climate.gov/news-features/understanding-climate/climate-change-atmospheric-carbon-dioxide
- Composition of greenhouse gas emissions https://www.epa.gov/ghgemissions/global-greenhouse-gas-emissions-data
- Carbon cycle diagrams – multiple options available via web search "Basics of Carbon Cycle and the Greenhouse Effect" https://www.esrl.noaa.gov/gmd/education/carbon_toolkit/basics.html
- Copy of Keeling Curve https://scripps.ucsd.edu/programs/keelingcurve/ http://www.esrl.noaa.gov/gmd/ccgg/trends/graph.html
- Diagrams including pre-Industrial Revolution http://www.ems.psu.edu/~radovic/COxTask3A.html
- Explanations of Keeling Curve available via web search Example: https://www.britannica.com/science/Keeling-Curve

Instructions for building diagrams and models

- From Cornell University: http://www.geo.cornell.edu/eas/education/course/descr/EAS302/302_06Lab11.pdf
- Carbon Cycle Model: http://mvhs.shodor.org/mvhsproj/carbon/carbon.html
- Video of a bathtub simulation https://www.youtube.com/watch?v=7Nt0v4YAAVg&t=57s
- En-ROADS climate simulator https://www.climateinteractive.org/tools/en-roads/

3.5 - The Greenhouse Effect

Media about the greenhouse effect

- Article about greenhouse effect https://climatekids.nasa.gov/greenhouse-effect/
- Video "The Greenhouse Effect" from US Environmental Protection Agency https://www.youtube.com/watch?v=VYMjSule0Bw
- Audio about methane as a greenhouse gas http://www.npr.org/templates/story/story.php?storyId=14288215 Scientist measures methane escaping from lakes in the far northern regions.

3.6 Global Warming

Historical behavior over time graphs displaying temperature and CO_2 levels

- http://climate.nasa.gov/
- https://earthobservatory.nasa.gov/Features/GISSTemperature/giss_temperature2.php

Introductory videos about global warming

- Video (Optional) "An Inconvenient Truth" 2006 (Full length about 1.5 hours)
  Sequel released 2017
- Five minute summary of global warming by Al Gore
  https://www.youtube.com/watch?v=jOKqYtRDGak
- "Scientist Measures an Overlooked Greenhouse Gas" NPR article and video
  Also linked in Lesson 3.5 The Greenhouse Effect
  https://www.npr.org/templates/story/story.php?storyId=14288215

Articles about effects of global warming

- "The Top Ten Global Warming 'Skeptic' Arguments Answered"
  https://www.theguardian.com/environment/climate-consensus-97-per-cent/2014/may/06/top-ten-global-warming-skeptic-arguments-debunked
- "It's Not Your Imagination. Summers Are Getting Hotter"
  https://www.nytimes.com/interactive/2017/07/28/climate/more-frequent-extreme-summer-heat.html
- "2018 is Shaping Up to be the Fourth-Hottest Year. Yet We're Still Not Prepared for Global Warming."
  https://www.nytimes.com/2018/08/09/climate/summer-heat-global-warming.html
- "2020 Tied for Warmest Year on Record, NASA Analysis Shows"
  https://www.nasa.gov/press-release/2020-tied-for-warmest-year-on-record-nasa-analysis-shows
- "A Historic Heat Wave Roasts Siberia"
  https://www.nytimes.com/2020/06/25/world/europe/siberia-heat-wave-climate-change.html
- "Siberian heatwave of 2020 almost impossible without climate change"
  https://www.worldweatherattribution.org/siberian-heatwave-of-2020-almost-impossible-without-climate-change/
- "How We Know It Was Climate Change"
  https://www.nytimes.com/2017/12/29/opinion/sunday/climate-change-global-warming.html
  Scientist connects destructive weather events to climate change. Includes behavior over time graphs.
- "Arctic Sea Ice Missed a Record Low This Winter. Barely."
  https://www.nytimes.com/interactive/2018/03/23/climate/arctic-ice-maximum.html
- "Current State of Sea Ice Cover"
  https://earth.gsfc.nasa.gov/cryo/data/current-state-sea-ice-cover
  Descriptions, graphs, and maps of Arctic sea ice
- "As Earth Warms the Diseases That May Lie Within Permafrost Become a Bigger Worry"
  https://www.scientificamerican.com/article/as-earth-warms-the-diseases-that-may-lie-within-permafrost-become-a-bigger-worry/
- "Portugal Forest Fires Worsen, Fed by Poor Choices and Inaction"
  https://www.nytimes.com/2017/08/12/world/europe/portugal-forest-fires-pedrogao-grande.html
- "How to live with mega-fires? Portugal's feral forests may hold the secret"
  https://www.nationalgeographic.com/science/article/how-to-live-with-mega-fires-portugal-forests-may-hold-secret
  Long, informative article about how changing climate and human activity contributes to fire intensity.
- "How Global Warming Fueled Extreme Weather Events"
  https://www.nytimes.com/2017/12/14/climate/climate-extreme-weather-attribution.html
  2016 was the hottest year recorded to date. Scientists linked these five extreme weather events

(among others) to global warming:
- ◊ record temperatures around the world
- ◊ coral bleaching
- ◊ drought in Africa
- ◊ wildfires in North America
- ◊ warm 'blob' in the Pacific Ocean

- "Wild Weather"
  https://www.nationalgeographic.com/climate-change/how-to-live-with-it/weather.html
  How extreme weather events are affecting humans.
- "In a Warming California, a Future of More Fire"
  https://www.nytimes.com/2017/12/07/climate/california-fires-warming.html
- "What climate change means for California"
  https://www.epa.gov/sites/default/files/2016-09/documents/climate-change-ca.pdf
- "Climate Change is Fueling Wildfires Nationwide, New Report Warns"
  https://www.nytimes.com/interactive/2018/11/27/climate/wildfire-global-warming.html
- "In Peru's Deserts, Melting Glaciers Are a Godsend (Until They're Gone)"
  https://www.nytimes.com/2017/11/26/world/americas/peru-climate-change.html
  Melting glaciers have been providing water for farming in arid regions. But the glaciers may soon be gone.
- "Vanishing Glaciers: The Future of Water in Peru's High Andes"
  https://news.climate.columbia.edu/2018/06/12/vanishing-glaciers-future-water-perus-high-andes/
  Retreating glaciers reduce water supplies for people living in the Andes.
- "What Climate Scientists Want you to See in the Floodwaters"
  https://www.nytimes.com/2017/09/02/opinion/climate-hurricanes-flooding-harvey.html
  Scientists explain the increasing number and size of floods and describe how people must adjust to them.
- "Extreme flooding to increase as temperatures rise, study finds"
  https://www.washingtonpost.com/weather/2021/09/13/extreme-flooding-increase-climate-change/

3.7 Examining Climate Models

Climate background info

- Gapminder home page
  http://www.gapminder.org/
- EPA webpages
  - ◊ https://www.epa.gov/climate-indicators
  - ◊ https://science2017.globalchange.gov/chapter/executive-summary/
  - ◊ http://scenarios.globalchange.gov/

Climate Interactive simulation website

Be sure to register and familiarize yourself with the program ahead of time.

- Climate Interactive home page
  https://www.climateinteractive.org/
- Climate Interactive Climate Summit
  https://www.climateinteractive.org/world-climate-simulation/

Chapter Four: Water

4.1 Protecting Water

Articles or research material about water

Drying waterways

- "In a Warming West, the Rio Grande is Drying Up"
  https://www.nytimes.com/interactive/2018/05/24/climate/dry-rio-grande.html

Drought reduces water available to people who depend upon the Rio Grande River.

- "Diverted, drained, and dwindling: What's the fate of New Mexico's Rio Grande?
https://www.hcn.org/articles/climate-change-diverted-drained-and-dwindling-whats-the-fate-of-new-mexicos-rio-grande
Drought and demand for water cause the river to dry up.

Products that pollute

- "Is Your Sunscreen Poisoning the Ocean?"
https://www.nytimes.com/2017/08/19/opinion/sunday/sunscreen-poisoning-ocean-reef.html
Chemicals toxic to coral threaten reefs' survival.
- "Sunscreen pollution accelerating demise of coral reefs, experts say"
https://abcnews.go.com/International/sunscreen-pollution-accelerating-demise-coral-reefs-experts/story?id=68807099
Includes video after advertisement.
- "Fertilizers, a Boon to Agriculture, Pose Growing Threat to US Waterways"
https://www.nytimes.com/2017/07/27/climate/nitrogen-fertilizers-climate-change-pollution-waterways-global-warming.html
Nitrogen runoff from agricultural fertilizers harms aquatic ecosystems.
- "Nitrogen and water"
https://www.usgs.gov/special-topics/water-science-school/science/nitrogen-and-water
- "Phosphorus and water"
https://www.usgs.gov/special-topics/water-science-school/science/phosphorus-and-water
Summaries of the effects of nutrients entering waterways
- "Planning for a Rainy Day"
https://www.nature.org/en-us/magazine/magazine-articles/planning-for-a-rainy-day/
How Washington, DC, works to reduce rainwater runoff from paved surfaces.

Plastics

- "The 'Great Pacific Garbage Patch' is Ballooning, 87,000 Tons of Plastic and Counting"
https://www.nytimes.com/2018/03/22/climate/great-pacific-garbage-patch.html
- "The Immense Eternal Footprint Humanity Leaves on Earth: Plastics"
https://www.nytimes.com/2017/07/19/climate/plastic-pollution-study-science-advances.html
Scientists calculate the amount of plastic humans have produced, most of it not in use but still around.
- "Small Plastic, Big Problem"
https://www.npca.org/articles/1862-small-plastic-big-problem
The prevalence of microplastics is a problem in all ecosystems.
- "Garbage Patches"
https://marinedebris.noaa.gov/info/patch.html
- "The world's plastic pollution crisis explained"
https://www.nationalgeographic.com/environment/article/plastic-pollution

4.2 Mosquito Nets and Overfishing

Article and video about malaria and mosquito nets

- "Meant to Keep Malaria Out, Mosquito Nets are used to Haul Fish In."
http://www.nytimes.com/2015/01/25/world/africa/mosquito-nets-for-malaria-spawn-new-epidemic-overfishing.html
Mosquito nets can protect people from contracting malaria, but using them for fishing has serious unintended consequences.
- "Mosquito nets: Are they catching more fishes than insects?"
https://www.sciencedaily.com/releases/2019/11/191111100910.htm

George Richardson's systems diagrams

- http://static.clexchange.org/ftp/newsletter/CLEx24.4.pdf

4.3 Sea Level Change

Research material about sea level

- Behavior over time graphs
  https://climate.nasa.gov/vital-signs/sea-level/

Background information

- "Is Sea Level Rising?"
  https://oceanservice.noaa.gov/facts/sealevel.html
- "Sea Level Rise, Explained"
  https://www.nationalgeographic.com/environment/global-warming/sea-level-rise/

Articles about sea level rise

- "Greenland's Melting Ice Nears a 'Tipping Point,' Scientists Say"
  https://www.nytimes.com/2019/01/21/climate/greenland-ice.html
  Accelerating melting of Greenland ice can be speeding up sea level rise.
- "Greenland Ice Sheet Reached Tipping Point 20 Years Ago, New Study Finds"
  https://news.climate.columbia.edu/2020/09/02/greenland-tipping-point-20-years-ago/
- "Jakarta is sinking so fast, it could end up underwater"
  https://www.nytimes.com/interactive/2017/12/21/world/asia/jakarta-sinking-climate.html
  Indonesian city of Jakarta is sinking, adding to troubles brought by climate-caused sea level rise.
- "Jakarta, the fastest-sinking city in the world"
  https://www.bbc.com/news/world-asia-44636934
- "A Crisis Right Now: San Francisco and Manila Face Rising Seas"
  https://www.nytimes.com/interactive/2020/02/13/climate/manila-san-francisco-sea-level-rise.html
  Two modern cities face crises caused by sea level rise.
- In Manila and San Francisco, rising seas are a crisis now"
  https://news.abs-cbn.com/news/02/14/20/in-manila-and-san-francisco-rising-seas-are-a-crisis-right-now
- "Explainer: Nine 'tipping points' that could be triggered by climate change"
  https://www.carbonbrief.org/explainer-nine-tipping-points-that-could-be-triggered-by-climate-change

Multimedia and links to videos

- "Earth's Albedo and Global Warming"
  https://mass.pbslearningmedia.org/resource/ipy07.sci.ess.watcyc.albedo/earths-albedo-and-global-warming/

4.4 Altering the Ocean Temperature and Chemistry

Part One: Cooling water experiment

- Lesson 4 from *The Shape of Change*, "It's Cool."
  http://static.clexchange.org/ftp/documents/x curricular/CC2010-11Shape4ItsCoolSF.pdf
- For building a computer model, download the following link and open the pdf file
  http://static.clexchange.org/ftp/documents/science/SC1999-09ItsCool.zip

Part Two: Information about length of day and temperature

- https://www.timeanddate.com/sun/
- https://www.almanac.com/astronomy/sun-rise-and-set
- http://www.usclimatedata.com/
- http://www.weatherbase.com/
- https://www.seatemperature.org/

Part Three: Suggested articles and resources about ocean heat and chemistry

Water's ability to store heat

- "Heat Capacity and Water"
  https://www.usgs.gov/special-topic/water-science-school/science/heat-capacity-and-water?qt-

science_center_objects=0#qt-science_center_objects

Overall threats to oceans

- "The World's Oceans are in Danger, Major Climate Change Report Warns"
  https://www.nytimes.com/2019/09/25/climate/climate-change-oceans-united-nations.html
  A United Nations report from September 2019 states that climate change is causing damaging storms, harm to ecosystems and fish stocks, and serious risk to coastal populations.
- "Climate Change Poses a Threat to Our Oceans"
  https://www.un.org/en/chronicle/article/climate-change-poses-threat-our-oceans

Ocean warming and climate

- NOAA resource
  http://oceanexplorer.noaa.gov/facts/climate.html
- "Ocean Warming is Accelerating Faster Than Thought, New Research Finds"
  https://www.nytimes.com/2019/01/10/climate/ocean-warming-climate-change.html
  Research cites the increasing rate of ocean warming and warns of catastrophic consequences.
- "Ocean Warming"
  https://www.whoi.edu/know-your-ocean/ocean-topics/climate-ocean/ocean-warming/

Ocean's role in the carbon cycle

- https://www.noaa.gov/education/resource-collections/climate/carbon-cycle

Ocean acidification

- Ocean Acidification information from Smithsonian Institution (includes videos)
  https://ocean.si.edu/ocean-life/invertebrates/ocean-acidification
- NOAA resources
  ◊ https://www.pmel.noaa.gov/co2/story/What+is+Ocean+Acidification%3F
  ◊ http://www.pmel.noaa.gov/co2/story/Ocean+Acidification
  ◊ https://www.noaa.gov/education/resource-collections/ocean-coasts-education-resources/ocean-acidification
- Woods Hole Oceanographic Institution
  http://www.whoi.edu/main/topic/ocean-acidification
- "Untangling the Many Ways Acidification Harms Marine Life"
  http://theyearsproject.com/learn/news/untangling-the-many-ways-ocean-acidification-harms-marine-life/

Coral reef conditions

- "Global Warming's Toll on Coral Reefs: As If They're 'Ravaged by War' "
  https://www.nytimes.com/2018/01/04/climate/coral-reefs-bleaching.html
- "How does climate change affect coral reefs?"
  https://oceanservice.noaa.gov/facts/coralreef-climate.html
- "Building a Better Coral Reef"
  https://www.nytimes.com/2017/09/20/climate/coral-great-barrier-reef.html
  Scientists seek ways to strengthen and restore coral reefs.
- "The Science of Super Corals" (video)
  https://oceantoday.noaa.gov/fullmoon-scienceofsupercorals/welcome.html
  Scientists assist corals to adapt to higher ocean temperatures.

How whales sequester carbon

- "How Whales Help Cool the Earth"
  https://www.bbc.com/future/article/20210119-why-saving-whales-can-help-fight-climate-change

4.5 Running a Fishing Company

Online Fishbanks simulation with related resources

- Creative Learning Exchange downloadable set of materials, links, and instructions (Highly recommended) written by Anne LaVigne, an outstanding educator and curriculum developer. The simulation may be accessed directly from MIT, but Ms. LaVigne's work will add value and guidance.
- Fishbanks packet from Creative Learning Exchange

- http://www.clexchange.org/curriculum/simulations/
  Scroll down to "Fishbanks."
- Simulation (link also available in Creative Learning Exchange packet)
  https://mitsloan.mit.edu/LearningEdge/simulations/fishbanks/Pages/fish-banks.aspx
- Article by Donella Meadows "The Tragedy of the Commons on Georges Bank, and Elsewhere"
  http://www.donellameadows.org/archives/the-tragedy-of-the-commons-on-georges-bank-and-elsewhere/

Reference Article on drawing stock/flow diagrams

- "Building the Fish Banks Model and Renewable Resource Depletion"
  http://ocw.mit.edu/courses/sloan-school-of-management/15-988-system-dynamics-self-study-fall-1998-spring-1999/readings/fish.pdf

Chapter Five: Pulling it All Together

5.1 Connecting Systems

Reference Articles

- "What a Week's Disasters Tell Us About Climate and the Pandemic"
  https://www.nytimes.com/2020/05/23/climate/climate-change-coronavirus.html
  A journalist explains how global climate and health crises coming together make difficult lives even harder.
- "Corona Virus and Climate Change"
  https://www.hsph.harvard.edu/c-change/subtopics/coronavirus-and-climate-change/
  Scientist answers questions about the connections between climate and public health.
- "Social Dimensions of Climate Change"
  https://www.worldbank.org/en/topic/social-dimensions-of-climate-change#1
- "Learn About Environmental Justice"
  https://www.epa.gov/environmentaljustice/learn-about-environmental-justice
- "Why 'I Can't Breathe' is Resonating with Environmental Justice Activists"
  https://www.nbcnews.com/science/environment/why-i-can-t-breathe-resonating-environmental-justice-activists-n1228561
  Racial prejudice extends throughout society, including issues of environmental justice.
- "Mold, Possums and Pools of Sewage: No One Should Have to Live Like This"
  https://www.nytimes.com/2020/11/14/opinion/sunday/coronavirus-poverty-us.html
  Poverty, inequality, illness, and environmental injustice in Alabama. (Note:graphic and disturbing)

Lesson 5.2 Choosing a Topic

No additional resources

Appendix C: Systems Resources

Systems Principles

When struggling to understand complex problems, sometimes it can lead to a feeling of déjà vu, a sense that we've been here before. Chances are a systems principle is at work. Systems principles are key ideas that repeat across time and within multiple contexts. This list is by no means comprehensive but provides a starting point for considering systems principles at play within a problem or issue of concern.

Everything is connected – there is no 'away."
Everyone uses mental models to make decisions.
Nothing grows forever – systems tend toward equilibrium.
Systems are usually resilient, resisting change but may still be pushed out of equilibrium.
Feedback is embedded in all actions.
Cause and effect are often not closely related in time or space.
There are seldom simple solutions.
Solutions often become problems.
Actions may be ineffective due to low leverage policies or incorrect application of high leverage ones.
What appears to be good in the short term is often bad in the long term, and vice versa.
Almost nothing is exogenous – causes of problems are within systems (endogenous).
Systems and problems within them may display parallel structures even though the contexts are unalike. (Generic structures)

A Summary of Systems Tools

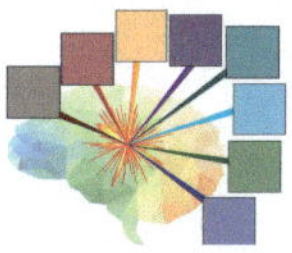

Barry Richmond's Systems Thinking Skills

Key question: How might looking at a system with different lenses help increase understanding and improve desired outcomes?

Use: This set of thinking skills provides a framework for looking at any situation, whether current, theoretical, or historical. Practice the skills one at a time or in concert with one another.

Behavior over time graphs (BOTGs)

Key question: What variables are changing over time?

Use: BOTGs are usually a good starting point. Find data where possible and be clear when graphs represent mental models versus research findings.

Connection Circles (CCs)

Key question: What elements in a system are affecting change in other elements directly?

Use: CCs are used for identifying important elements in a system and how they affect each other. Tracing causal connections can help find feedback loops.

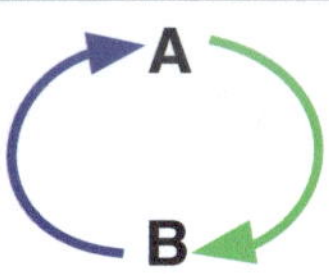

Causal Loop Diagrams (CLDs)

Key question: Where are feedback loops in the system?

Use: Complex systems usually have "competing loops" – try to find reinforcing loops and balancing loops that may be in conflict.

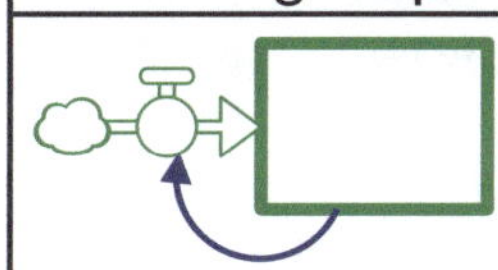

Stock/Flow Diagrams (S/Fs)

Key questions: What is accumulating or diminishing in the system? What are the rates of change?

Use: Stocks are changed only by flows and their units must be the same. Stock/flow diagrams show how things actually work.

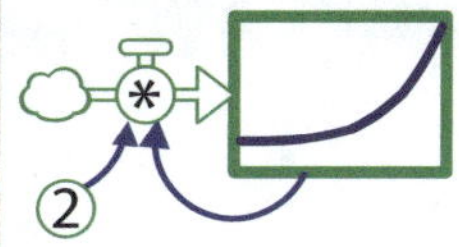

System Dynamics Models

Key question: How can we mathematically define interdependent elements within a system?

Use: Models help illustrate how systems really operate, showing results of their internal interconnections over time.

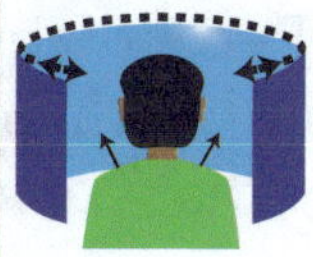

Simulations

Key question: How can we test assumptions and try out "what if …" strategies?

Use: Simulations help compress time and space, so rather than having to wait a hundred years to see a particular result, a simulator can reveal that result quickly and in a contained space, either real or virtual.

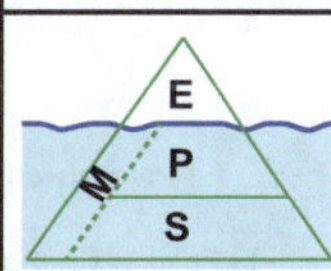

Iceberg Visual

Key question: Looking beyond easily seen events, what lies under the surface, often unseen?

Use: The iceberg provides a frame for looking beyond events to exploring a system more deeply - looking for trends, structures and mental models within. Because complex systems problems resist change, seeing a system through the lens of an iceberg can help reveal what's generating behaviors seen.

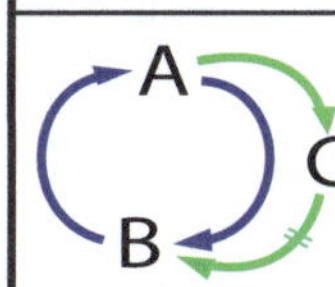

Archetypes and other Generic Models

Key question: Can we identify a generic structure, a pattern or behavior that occurs in a variety of situations?

Use: Archetypes (and other generic models) can help in the transfer of knowledge about a specific subject to other situations. Viewing behavior in a broader interdisciplinary context can reveal repeating patterns and increase understanding of potential outcomes that may be counterproductive to one's goals.

Barry Richmond's Systems Thinking Skills

Forest Thinking – As if viewing the system from 10,000 meters, seeing a big picture rather than focusing on details

Dynamic Thinking – Recognizing patterns rather than events, realizing that continuous processes rather than discrete events create the conditions in the real world – understanding that you can always quantify, though you can't always measure

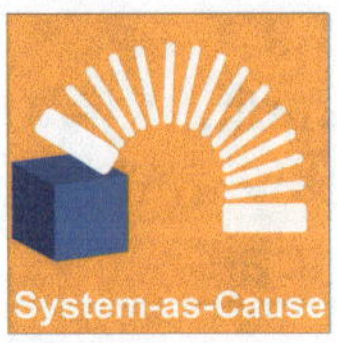

System as Cause Thinking – Placing responsibility for a behavior on internal actors, self-generating rather than external forces

Operational Thinking – Understanding how things really work, e.g., maintaining consistent units of measure, and expecting that factors don't always cause proportional effect

Closed-Loop Thinking –Tracing ongoing interdependent processes rather than one-way causal chains

Scientific Thinking – Testing hypotheses, applying rigorous simulation and quantification

Empathic Thinking – Giving and receiving feedback, expanding personal boundaries, avoid taking a single rigid position, and appreciating a spectrum of opinions

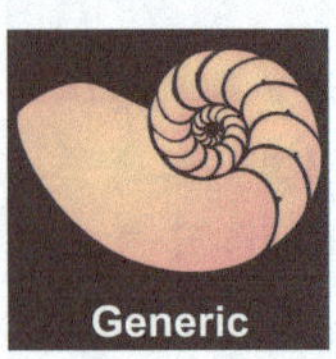

Generic Thinking – Recognizing generic structures rather than specific events

Adapted from Tracing Connections by Creative Learning Exchange, 2021

Introduction to Behavior Over Time Graphs (BOTGs)

Behavior over time graphs (BOTGs) are line graphs with time along the x-axis (horizontal) and the scale for one or more variables along the y-axis (vertical). Emphasize the overall shape of the change rather than the exact position of the data points.

1. Hard versus soft variables - It's the shape that matters.

   BOTGs are designed to represent our thinking. Some graphs display quantifiable or 'hard' variables. For example, scores on quizzes or number of homework assignments turned in. Other BOTGs represent 'soft' variables, such as confidence, satisfaction, or enjoyment. All BOTG graphs allow our 'mental models' to take a visual form so we can share them, or analyze them ourselves. Almost all BOTGs are more insightful for their general shape and trend rather than the actual data points. Are things moving in the right direction?

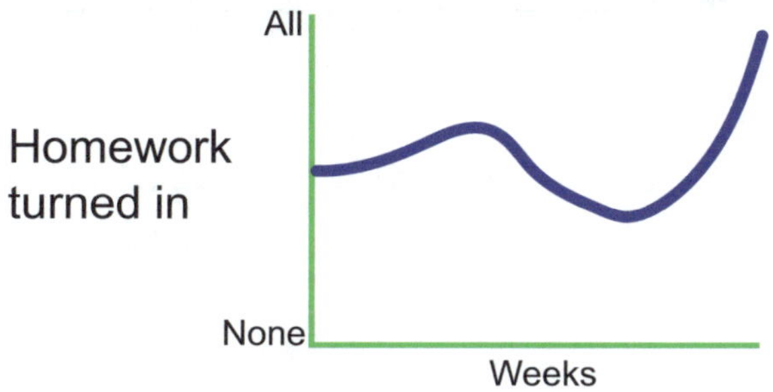

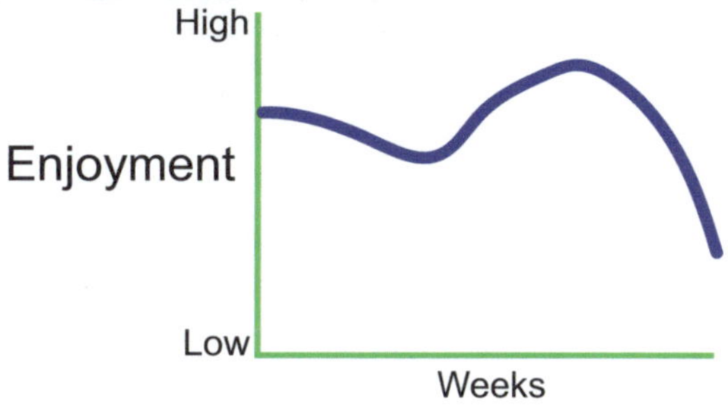

2. Two variables can be better than one.

   Adding a second variable on the same graph often increases the learning by an order of magnitude.

   For example, consider a student's performance graph that shows a fairly flat line for quiz scores over the first few weeks of school. Then the line rises, representing steady linear increase since November 1st. Add a second variable, like confidence, to the same graph pad, keeping in mind that the y-axis scale may have to be different for the second line.

   Two or more lines on one graph can help when considering whether those variables are causal, correlated, or coincidental.

3. Questions raised by BOTGs

   Reflect on the implications of the BOTGs using this simple structure:

   - What is changing? Identify the variable(s) being graphed and the story it tells.
   - How is it changing? Describe the trend, slope, or trajectory.
   - Why is it changing? Think about the causes and structures driving the behavior shown on the graph.
   - So what and now what? Evaluate the significance and consider what to do based on the information and insight gained.

See more about using BOTGS: www.clexchange.org/gettingstarted/botg.aspx

Behavior-over-time Graphs

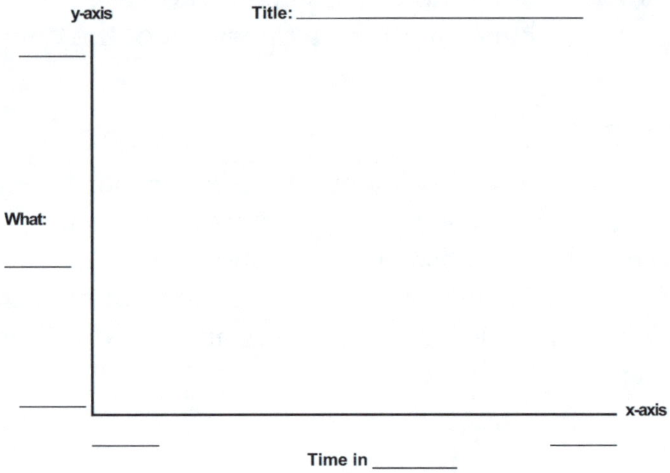

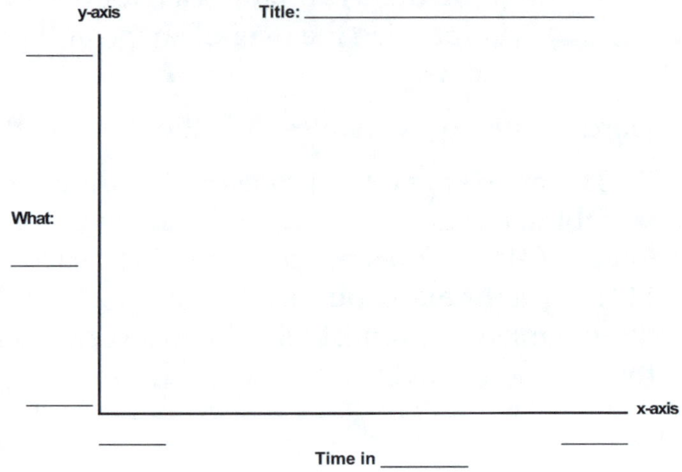

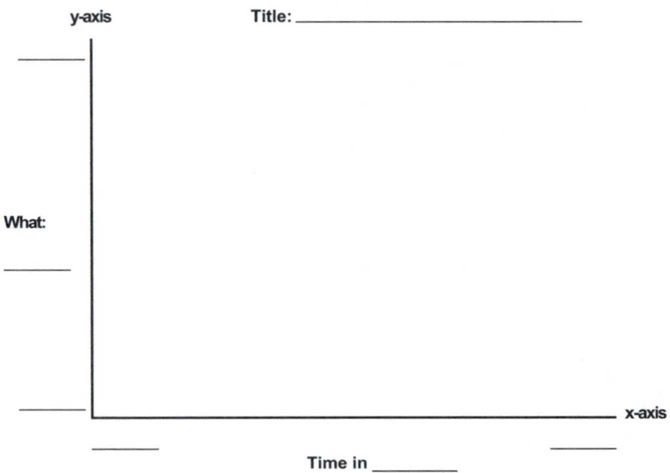

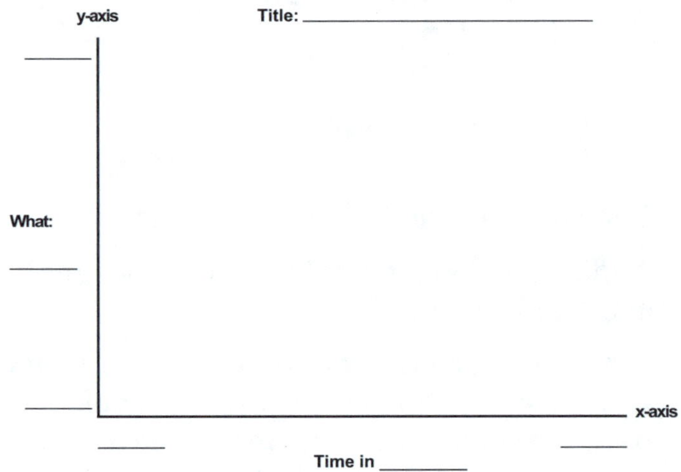

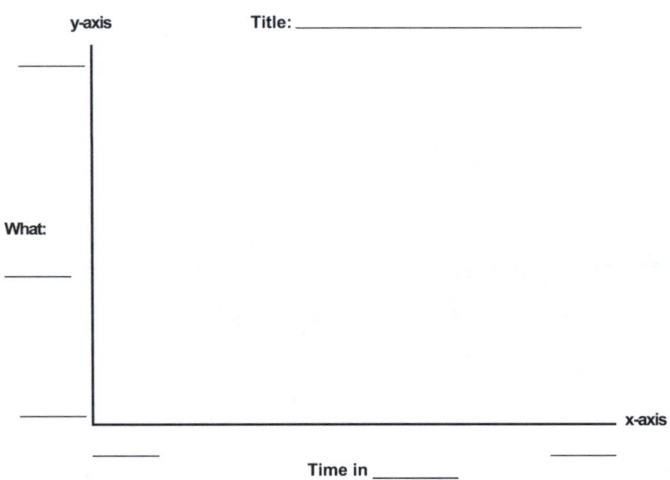

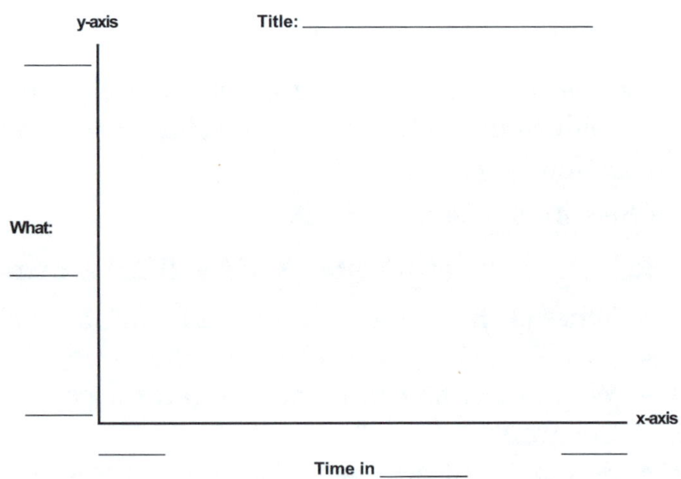

CREATIVE LEARNING Exchange Making Thinking Visible

Handout: Behavior Over Time

Create your graph and then answer the questions.

Title: ____________________

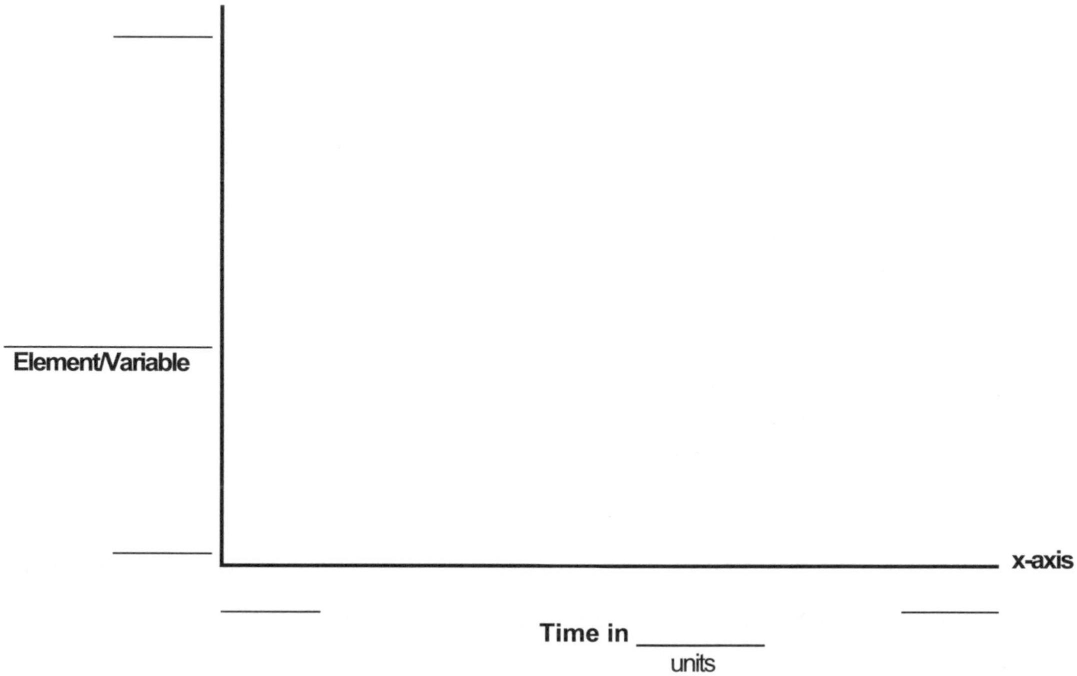

What is changing? Identify the variable(s) on the graph. Explain how the data was measured and other details that are important.	
How is it changing? Describe the rate of change shown by the shape of the graph.	
Why is it changing? Tell why you think (based on your research) the variables behaved as they did.	
So what and now what? What is the significance of the graph? Decide what policies or decisions these data might require.	

Introduction to Connection Circles (CCs)

A connection circle is a visual tool that shows the relationships between variables in a story or complex situation. When using a connection circle consider:

What are the key factors or variables in this story?

What are the cause and effect relationships between the variables?

Instructions:

1. Use a template or draw a circle on a piece of paper.
2. Choose elements from the story or situation that meet these criteria:
   - They are important to the changes in the story.
   - They are nouns or noun phrases.
   - They increase or decrease in the story.
3. Write your elements around the circle. (Select no more than 5–10)
4. Look for causality. Find elements that cause another element to increase or decrease.
   - Draw an arrow through the circle from the cause to the effect.
   - The causal connection must be direct.
   - Label the arrowhead with a "+"or a "-"

   A "+" (plus sign) indicates that a change in the first factor becoming greater or less causes the second element to change in the same direction.

   A "-" (minus sign) indicates that a change in the first element causes the second element to change in the opposite direction.

   You can use different colored pencils to indicate "+" vs. "-".
5. Find all possible causal connections. Some links may be based on actual data, and some may be simply a hypothesis. Try to draw at least one causal arrow into and out of each element.

In connection circles, you can trace closed loops of cause and effect, also called feedback loops or causal loop diagrams.

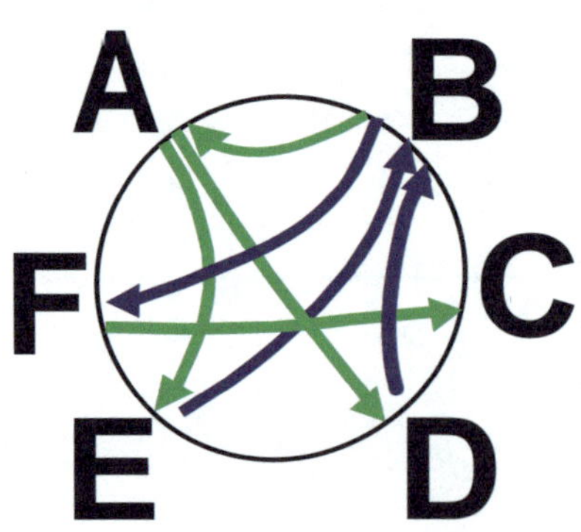

See more about using Connection Circles at www.clexchange.org/gettingstarted/connectioncircles.asp

Connection Circle Template with Guidelines

1. Draw a large circle and list 5-10 important elements (that can increase and decrease) around the circle.
2. Identify an element that causes another element to increase or decrease.
3. Draw an arrow through the circle from the cause to the effect. Make sure that the causal connection is a direct one.
4. Identify polarity (+ or –) of link and label at the arrowhead.
5. Continue to identify elements with causal connections.
6. Identify feedback loops and tell emerging "stories."

Introduction to Feedback Loops or Causal Loop Diagrams (CLDs)

"Feedback loops control a system's major dynamic behavior. A feedback loop is a series of connections causing output from one part to eventually influence input to that same part." (Jack Harich)

In systems thinking, which deals with changes over time and the causal elements that drive those changes, feedback is identified as a loop mechanism. Causal loop diagrams (CLDs) are visual representations of the two types of feedback loops.

Reinforcing (R) or positive loops.
Think about a performance improvement loop, or virtuous circle. Practice some skill, such as playing music or athletics, and we get better. When we get better, we can perform at a higher lever, and we can get even better. Unfortunately, vicious cycles exist also, like inflation. Prices go up, so workers negotiate higher wages. Higher wages increase buying power, but companies raise prices as it becomes more expensive to produce goods, partly because of increased wages. These up or down spirals are called reinforcing or positive loops.

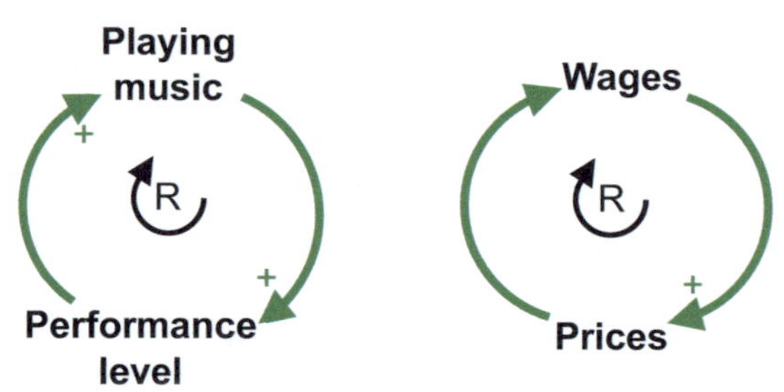

Balancing (B) or negative loops.
Let's look at performance in a different way. What if I don't care about getting better at my lessons? But my parents care and apply pressure to me. So I practice and I get better. Once I satisfy them, however, I can back off on practicing. Then my skills begin to erode. That makes my parents react, so I have to increase practice again. I stay with this oscillating pattern, practicing enough to keep my parents off my case, but not working so hard that I forego other activities I'd rather do. This "wave-like" pattern is a balancing loop. A thermostat is a good example of a balancing loop in our everyday life. Heating or cooling is called for when a certain temperature criterion is passed. Once it is met, the machine doing the work shuts off, until the measure of heat in the room turns it on again. Notice how the heating and cooling loops are similar and different.

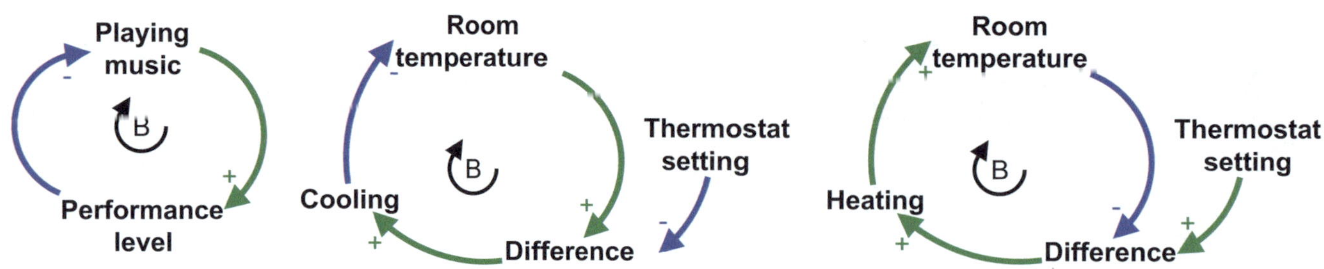

The examples above show two different mental models about practicing music and heating a room. They illustrate how using systems tools can help surface mental models and improve communication.

See more about causal loops at www.clexchange.org/gettingstarted/causalloops.asp

Introduction to Naming and Labeling

Naming Elements

1. Elements in systems diagrams need to go up and down.
2. Avoid quantitative adjectives. For instance, say "Population" rather than "More population."

Labeling Links (causal arrows) in CLDs and Stock/Flow Diagrams

1. Label each arrow at its head according to its polarity to indicate the direction of change
2. Be consistent and use either + and – or S(ame) and O(pposite)

   + or S means that a change in the first element causes a similar change in the second element or adds to the second element. For example,

   ◊ When the first becomes greater, the second becomes greater.
   ◊ When the first becomes smaller, the second becomes smaller.

   - or O means that a change in the first element causes an opposite change in the second element or subtracts from the second element. For example,

   ◊ When the first becomes greater, the second becomes smaller.
   ◊ When the first becomes smaller, the second becomes greater.

   + and – are used in the diagrams in this book.

   Note: Novices to systems thinking (both teachers and students) may find the S/O method of labeling to be easier than the +/- method as they initially explore connection circles and causal loops.
3. Causal arrows only have one head; they are one-way. When two elements affect each other, draw two different arrows.
4. Capitalize the first word of each element's name.

Tips for Labeling Links

TIP #1
To label a link, always start with the element at the beginning of the arrow going up. For 'Element A' → 'Element B', ask "If 'A' goes up, Does 'B' go up or down?" If 'B' goes up, label the arrowhead with a '+'; if 'B' goes down, label with a '-'. Do this same process to label each of the arrows independent of others. When all links are labeled, talk about the loop as a full story going all the way around.

TIP #2
If using the 'S' and 'O' method, an easy method for labeling links is using thumbs. Start at the origin and point one thumb the way that element is going and then go to the next element and use the other thumb to respond to the question, "When this goes up, does the other one go up or down?"

Continue around the loop until all links are labeled. Tell the story of the loop.

Labeling Causal Loops

Causal loops come in only two flavors, reinforcing and balancing (See Introduction to Feedback Loops..., p 220.) By labeling and naming causal loops, it creates useful reference points for analyzing and discussing systems.

Steps:

1. Determine loop type with one of two methods
   Option 1: Count link types

- Count the number of '-' links in any loop.
- An even number indicates the loop is reinforcing; an odd number indicates the loop is balancing.

Option 2: Tell the story

- Always start with any loop element going up.
- Tell the story of the loop going around until back to the original element.
- If it's still going up, the loop is reinforcing. If it's now going down, it's balancing.

2. In the center of the loop, place an 'R' if the loop is reinforcing or a 'B' if the loop is balancing.
3. Draw a circular arrow around the 'R' or 'B' to show the direction of the loop.
4. Add a word or short phrase in the middle that describes the essence of the loop.

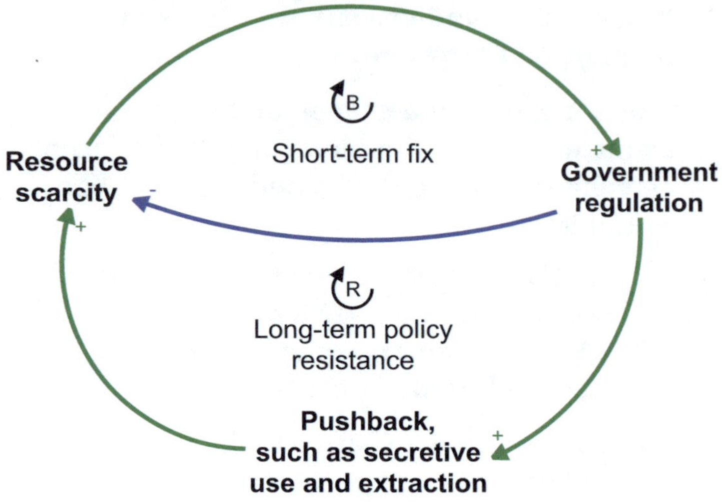

Labeling stock/flow diagrams

1. Stocks are accumulations. Label them as nouns or noun phrases.
2. Flows are the rates of change in the system. Labeling options include Gerunds (-ing words) e.g., Producing, Dying, Leaving, Growing, ...
3. Rates are an amount per time, e.g., Births per year, Widgets per month, ...
4. Be consistent with capitalization. In this book we capitalized the first word in the names of stocks, flows, and converters.
5. Be sure the units going into and out of stocks match the units of the stock.

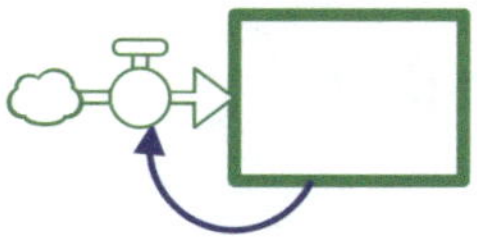

Introduction to Stock/Flow Diagrams (S/Fs)

A principle of systems thinking involves stocks, accumulations of things, and flows, measurement of the rates at which the stocks change. The clarity of stock/flow diagrams aids in appreciation for rates of change, feedback, and other elements of systems.

"Water in a bathtub" is an example of a stock. The flow from the faucet is an inflow that fills the tub, while the drain is an outflow to empty it.

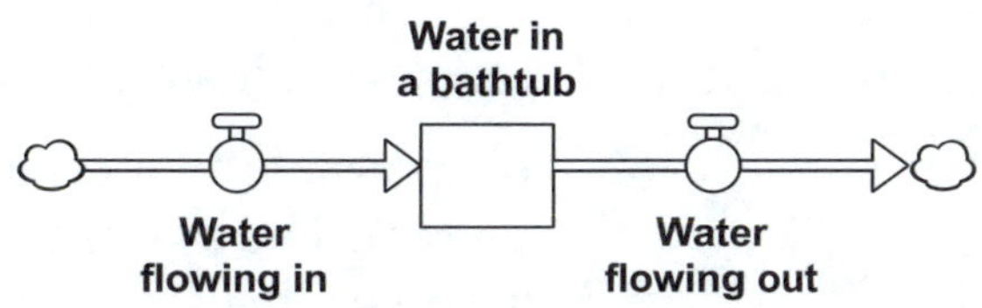

The clouds on the ends of the flows mark the model boundaries. It is important to know where water comes from to fill the tub and where it travels when it drains, but this diagram does not model the larger system. This is a simple bathtub diagram.

Substitution of other names for the elements in our diagram can give insight into other systems, e.g., "CO2 in the atmosphere":

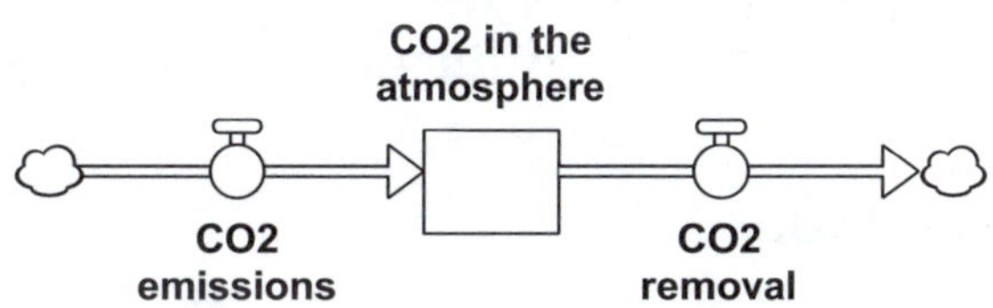

Most people agree this causes a big problem, climate change. If "CO_2 emissions" level off close to 2020 levels, the inflow will still exceed the outflow. And when inflow > outflow, the stock increases. That's true for anything we can represent as a stock and flow.

Another substitution shows insight about the use of credit cards:

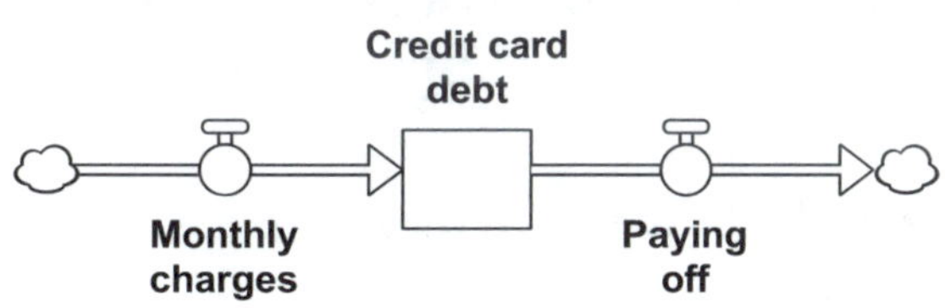

If spending is more than what is being paid off, the stock of debt rises. Debt costs money in the form of interest, and adding another element helps present a more complete picture.

"Monthly charges" are based on a rate represented by a circle called a converter. The "Interest rate" is multiplied by the existing debt, adding to the stock of "Credit card debt." This feedback connection is shown by the arrows called connectors. Students can easily learn how to input numbers and equations in order to model this situation.

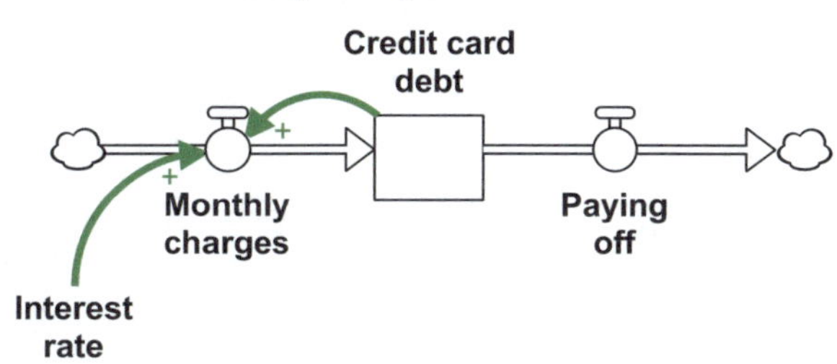

See more about stock/flow diagrams at www.clexchange.org/gettingstarted/stockflow.asp

Introduction to System Dynamics Computer Models

By creating system dynamics computer models, students of all ages can compress both space and time, thereby discovering long-term results of actions. Because modelers interact through computers and generate results in graphical, tabular, and animated formats, they maintain a high level of attention, engagement, and, therefore, learning. Even the simplest modeling of a system can reveal insights typically missed without it.

> NOTE: This process can feel overwhelming to a novice, so one option is to find simple models that others have created that relate to the issue of concern. Then using the other steps in the modeling process, test and/or modify those models as a way to explore the issue. For example, if concerned about famine in a particular country, look for a basic population/resource model.

It's possible to model any dynamic, interdependent system. The system dynamics modeling approach involves the development of computer simulation models that portray processes of accumulation and feedback and allow for the testing of "What if" questions. Systematically testing models can help with identification of effective strategies for overcoming a system's natural tendency to return to a status quo or equilibrium. (Think about how difficult it is to change an ingrained habit.)

Steps in the modeling process:

1. Start with a specific problem or issue.
2. Brainstorm a list of key variables related to the problem. Possibly use connection circles to pull out pertinent feedback loops.
3. Sketch behavior over time graphs (BOTGs) for key variables as possible.
4. Create an endogenous (meaning from within) stock/flow diagram of the problem, preferably collaborating with at least one other person. With models, it's important that the dynamics of the problem are generated from within as opposed to being generated from an outside influence.
5. Make sure that the elements in the system are interconnected with loops that show information feedback, i.e., circular causality.
6. Create mathematical equations to represent the relationships among elements in the system.
7. Run the model (asking "What if" questions) to test it with relevant data and information.
8. Create documentation (e.g., define variables, discuss key elements, explain equations) for the model.
9. Tell the "story" of the model as a way to communicate learning and insights.

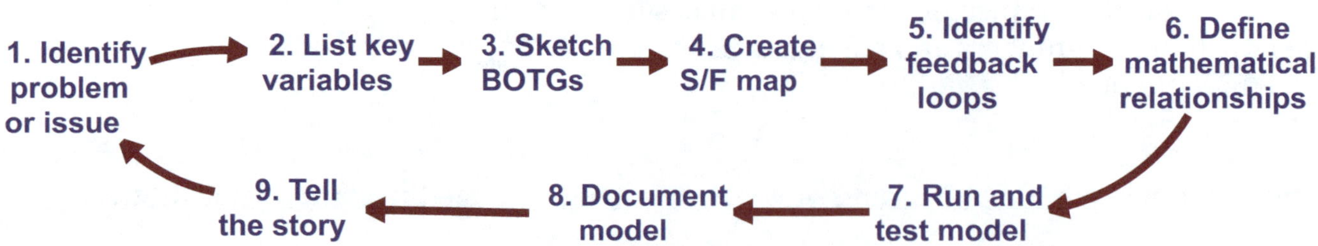

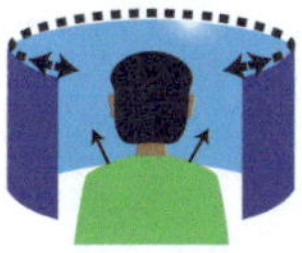

Introduction to Systems Simulations

Systems simulations are dynamic activities that include the accumulations (stocks) in a system in order to compress time and space. Simulations enhance the ability to discover long-term results of actions taken within a system. They allow the participant to become part of the system in order to explore "what if" questions as well as to make predictions and track dynamic behavior that results from the systemic structure.

Simulations actively engage participants, usually in cooperative situations. The fun, hands-on experiential nature of simulations tends to increase learning and encourages asking good questions. Several types of simulations can help students of all ages grasp the principles of systems as well as aid in exploration of individual systems.

<table>
<tr><th>Type</th><th>Example</th></tr>
<tr><td>Kinesthetic
Created with moving physical parts, such as people or objects that represent key accumulations of a system. Use anytime, and in particular, to scaffold understanding towards more abstract representations of systems.</td><td>
The In and Out Game with individuals walking into and out of a stock of People.</td></tr>
<tr><td>Physical
Aims to replicate aspects of a real world system, often in isolation or as a smaller version of the real one. Use as part of the scientific method to illustrate key behaviors and track data over time to later connect to understanding how the larger, real world system behaves.</td><td>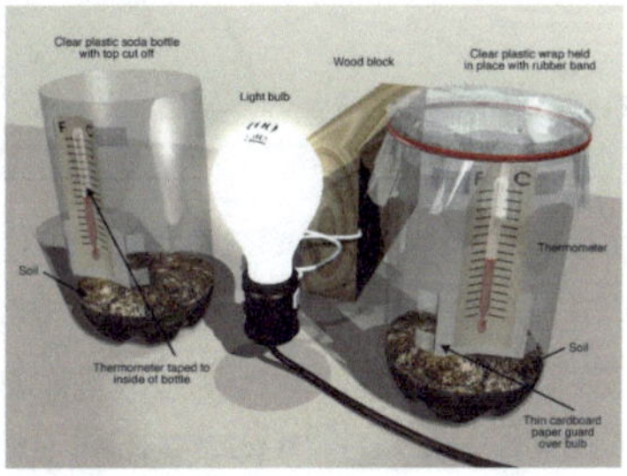

The creation of a mini-greenhouse out of easily obtained materials (Lesson 3.5)</td></tr>
<tr><td>Computer
Has an interface that overlays a system dynamics model, guiding users to explore how changing a model's assumptions can change resulting behavior. Use as a way to explore "What if" questions, especially when building one's own model is not possible or practical.</td><td>

Use a computer simulation to learn about Predator/ Prey relationships.</td></tr>
<tr><td>Computer and Physical
Combines elements of real and virtual representations of systems. Use to enhance understanding of a physical system when testing potential impacts of proposed policies.</td><td>
The World Climate Simulation from Climate Interactive allows participants to run a computer simulation to generate results with data from interactions among participants during the simulation. (Lesson 3.7)</td></tr>
</table>

To learn more and to find links to other available simulations, visit the Creative Learning Exchange at http://www.clexchange.org/curriculum/simulations/

Introduction to Iceberg Thinking

Beyond Events

Thinking beyond individual events to deeper elements of a system can help create more effective change. This is called iceberg thinking. Events, like an oil rig fire, are what we can typically observe at the top of an iceberg. These events are very important, but changing any individual moment is unlikely to create systemic change. To find real leverage, we must delve deeper to examine what we don't typically see, 90% of an iceberg (system) that's below the water's surface.

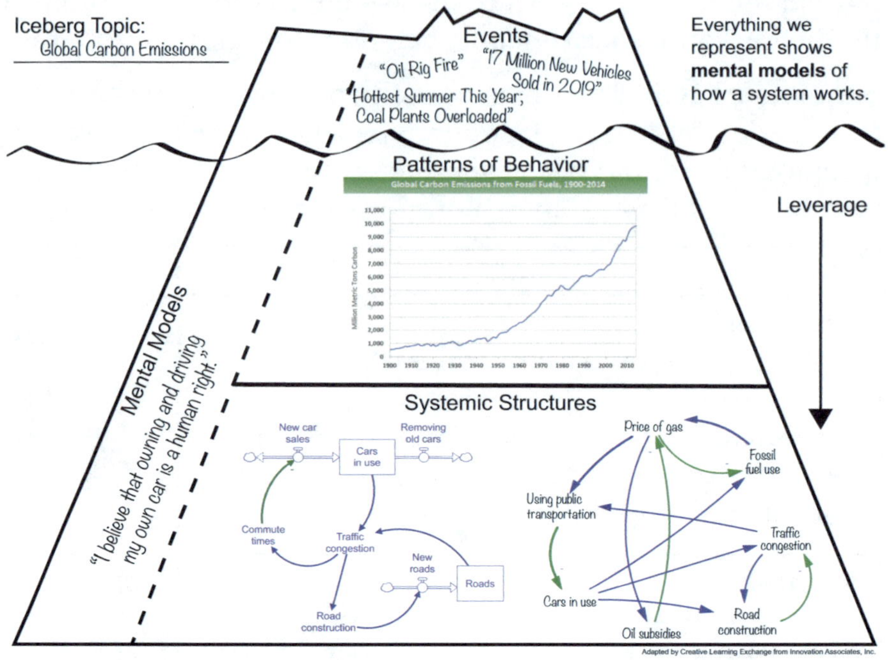

Patterns of Behavior

Below the surface are underlying patterns of behavior, i.e., trends. For example, "What is happening to carbon emissions over time?" or "How many cars are on the road over time?" Notice the common thread is the phrase "over time." It means weaving data points into a representation that shows an underlying trend. Notice the overall trend on the graph for global carbon emissions shows compounding growth.

Tool to Use: Behavior over time graph

Systemic Structures

Interconnected structures generate the trends that emerge. For example, if car sales are increasing over time, how does that affect road construction and fossil fuel use? How do policies and laws, e.g., subsidies for the fossil fuel industry, impact fossil fuel use in comparison to renewable energy use? The stock/flow and connection circle diagrams show some of these interdependencies with arrows that illustrate cause/effect relationships and feedback loops.

Tools to Use: Stock/flow diagram, connection circle, causal loop, simulation, system dynamics model

Mental Models

Peter Senge defines mental models as "... deeply ingrained assumptions, generalizations, or even pictures or images that influence how we understand the world and how we take action." There's a place on the iceberg diagram to make note of these mental models, but keep in mind that anytime we discuss or represent how systems work using systems tools, we are really surfacing our mental models. Talk with a dozen people about emissions from fossil fuels, and you'll surface at least a dozen different mental models.

Tools to Use: All systems tools

Iceberg Template

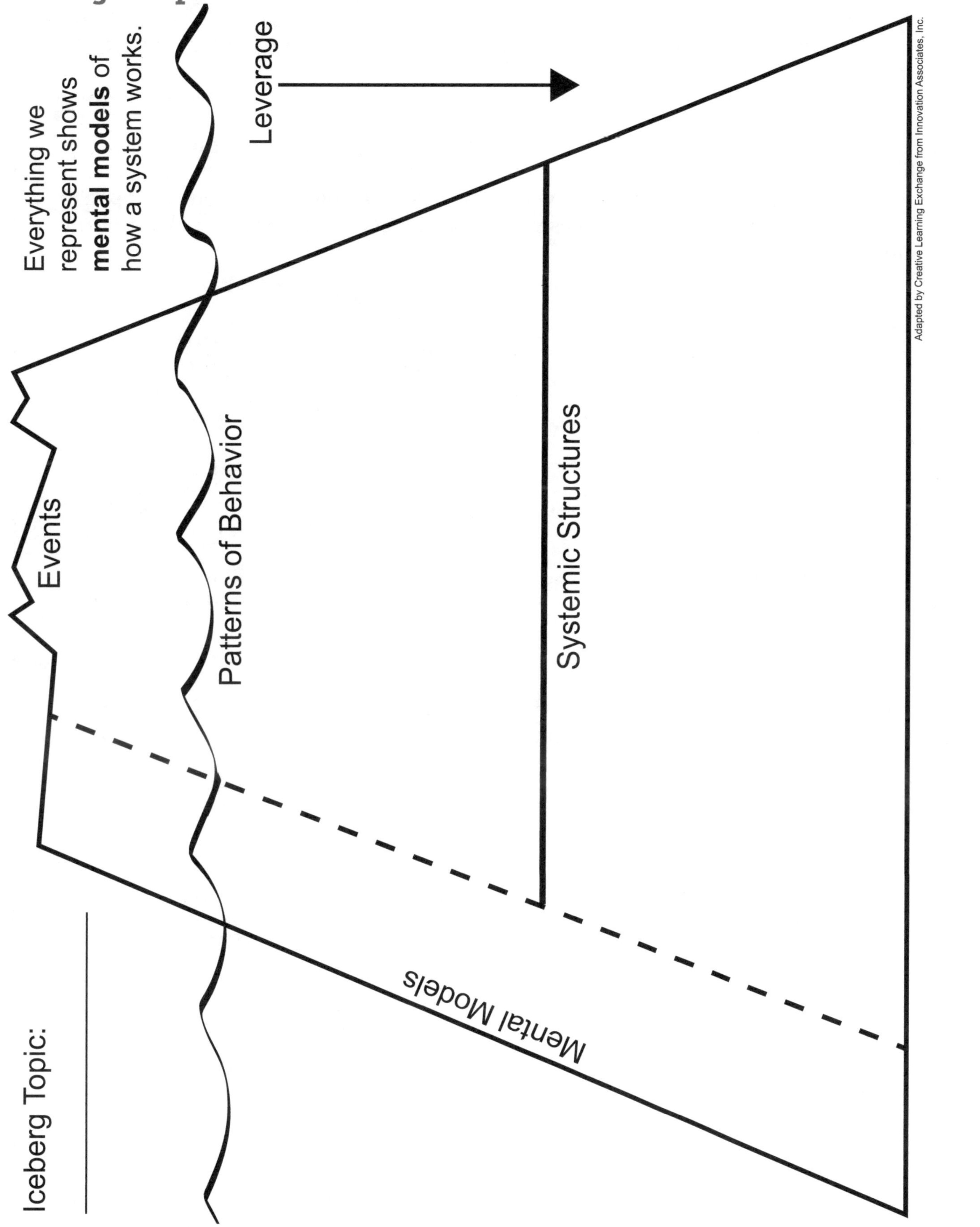

Introduction to Systems Archetypes (and Generic Models)

Systems archetypes are tools that capture common stories of dynamic behavior that repeat in diverse situations. Although others exist, these are particularly useful for the contexts in this book.

Tragedy of the Commons

When too many people exploit resources that are held in common, those resources are potentially used faster than they can regenerate (in the case of renewable resources) or disappear entirely (in the case of non-renewable resources). Example: As businesses open to serve both residents and tourists, it increases demand for services, which can put a strain on city infrastructure.

GENERIC ARCHETYPE

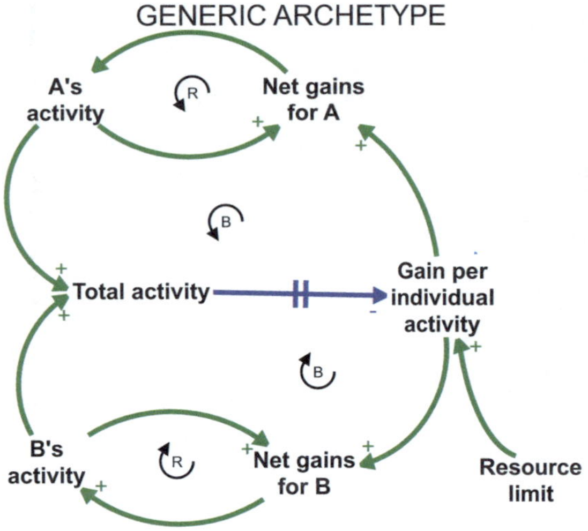

EXAMPLE

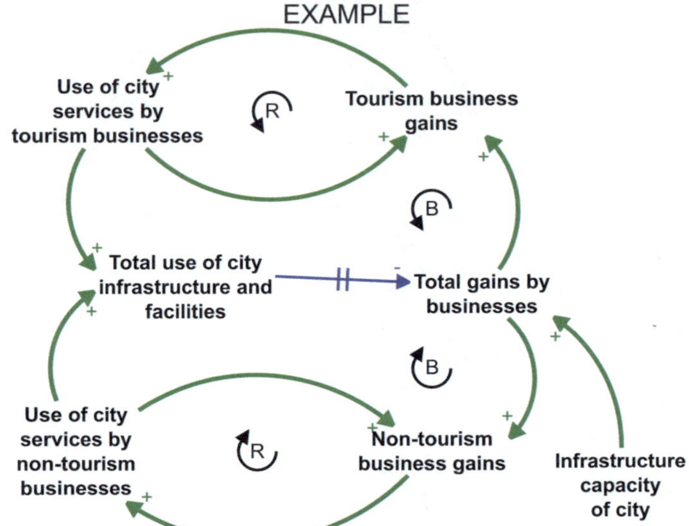

Escalation

Competition creates an "arms race" in which two entities regard the actions of the other as a threat. Example: Destructive behaviors among political parties can escalate as a way to maintain popularity and power.

GENERIC ARCHETYPE

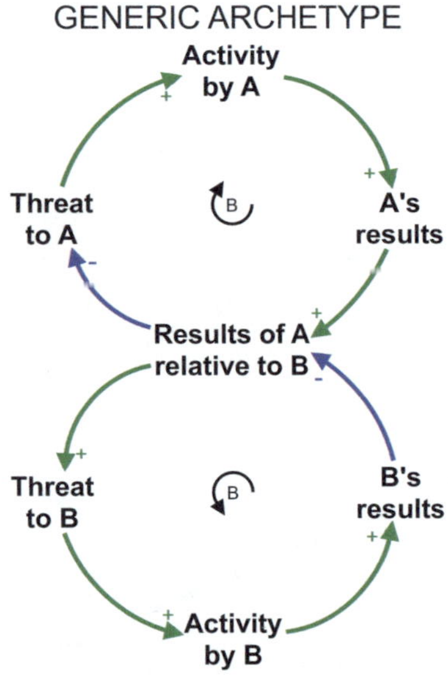

EXAMPLE

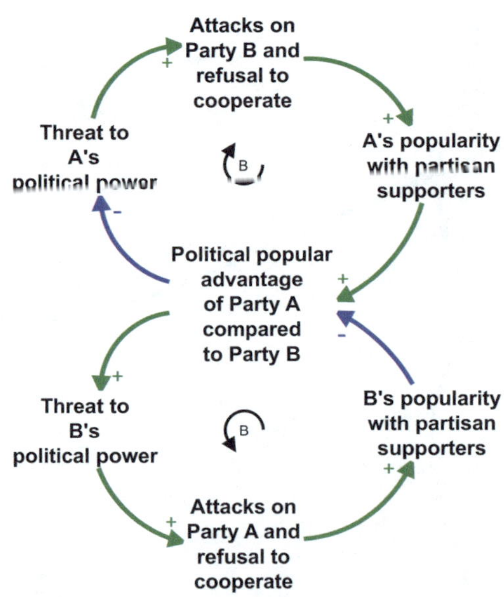

Drifting Goals (also called Eroding Goals)
A fundamental solution for reaching a goal is ignored or changed because of pressure to lower the goal. Example: The further we are from reaching a desired level for CO_2 concentration, the more pressure exists to lower that goal.

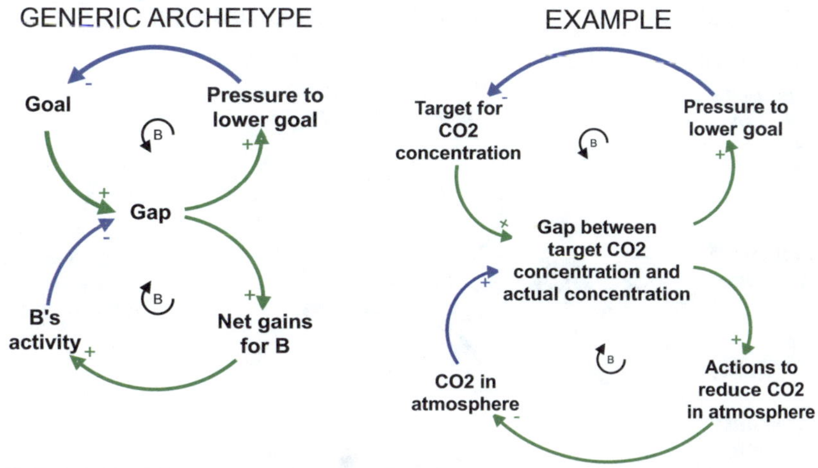

Fixes that Fail/Unintended Consequences
An apparent solution to a problem causes unintended consequences that make the situation worse. Example: Strict incarceration policies can backfire, leading to even higher levels of crime.

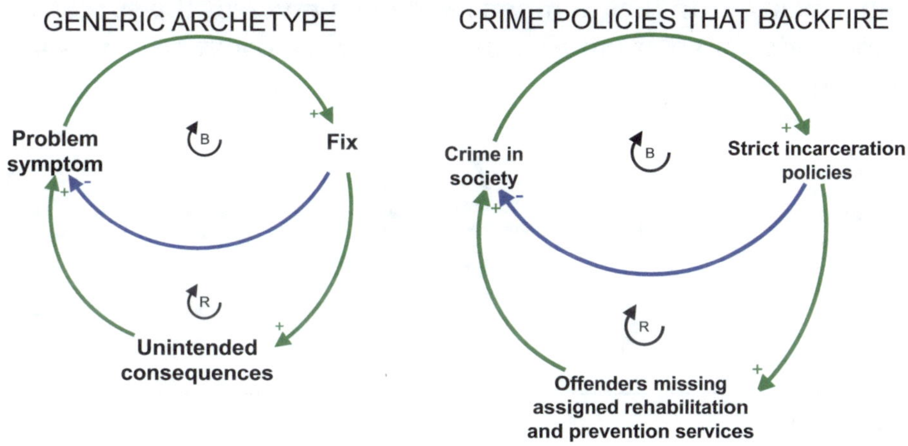

Shifting the Burden

A strategy is used to treat symptoms with a short-term solution in place of choosing a fundamental solution. The more a short-term solution is used, the greater the side effect is, thus diminishing the use of the more fundamental solution. Example: Government may reduce services as community organizations try to help needy populations.

GENERIC ARCHETYPE

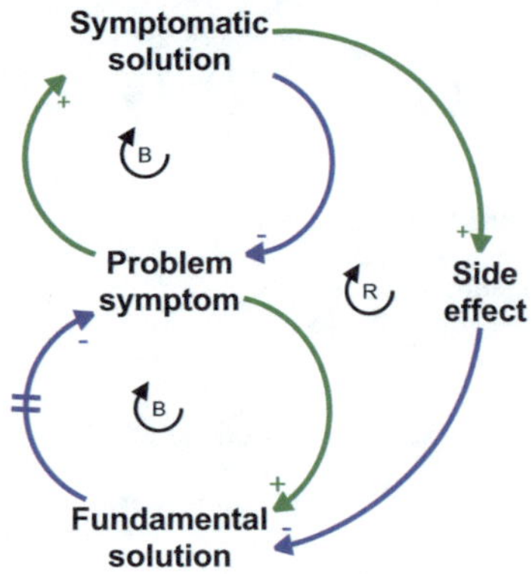

EXAMPLE

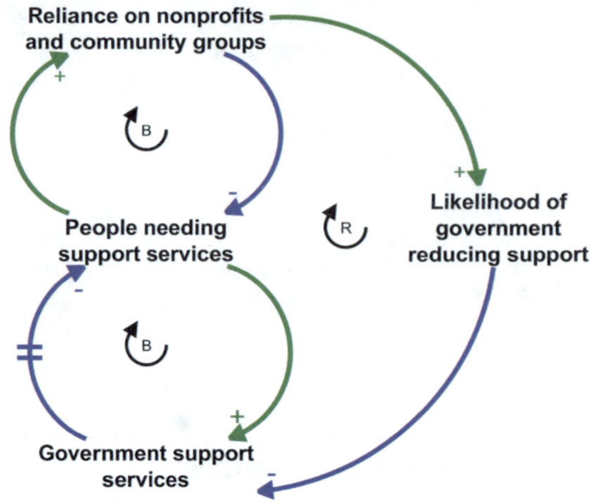

Limits to Growth (also called Limits to Success)

Growth can be rapid, building as a reinforcing loop. But encountering limiting factors will eventually create a balancing loop that dampens and even reverses the success. Example: Buildings that are not ready for solar panels can slow down a conversion to solar energy.

GENERIC ARCHETYPE

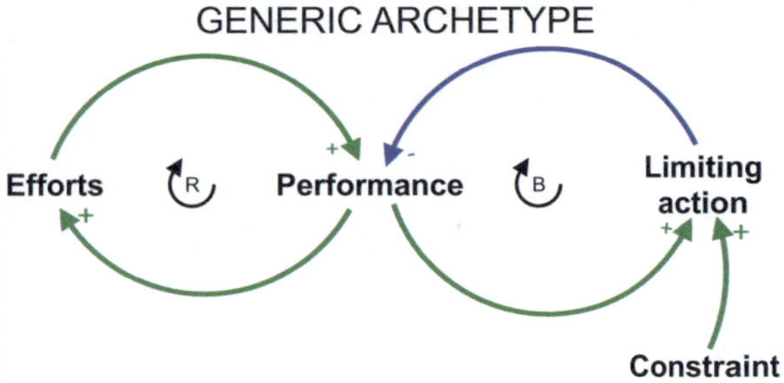

EXAMPLE

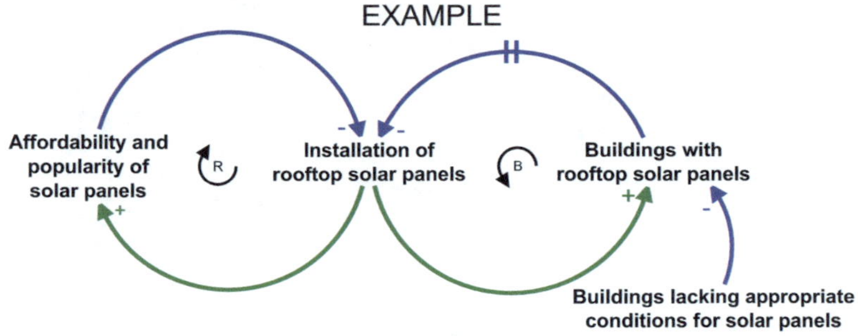

Success to the Successful

As one group is more successful than another, that group may receive more resources in comparison to another group that's struggling with success. Example: The fossil fuel industry has received subsidies, possibly reducing or eliminating subsidies for renewable energy conversion.

GENERIC ARCHETYPE

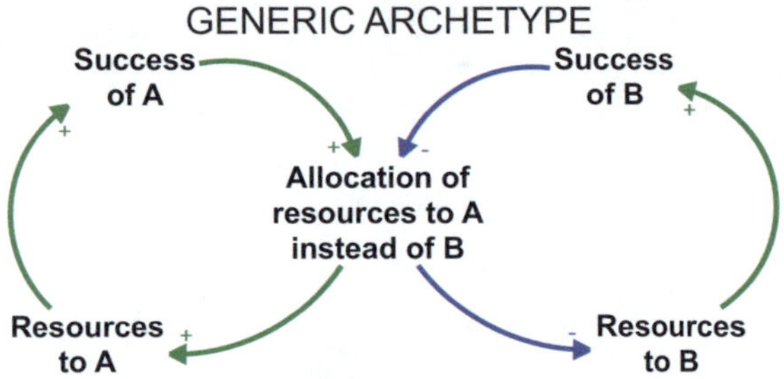

EXAMPLE

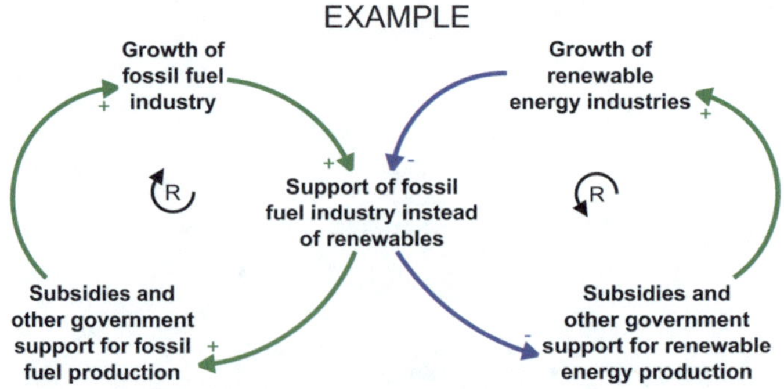

Template: Tragedy of the Commons

Use the guide at right as you complete your own example of this archetype below. Create a title and tell the "story" that explains your example.

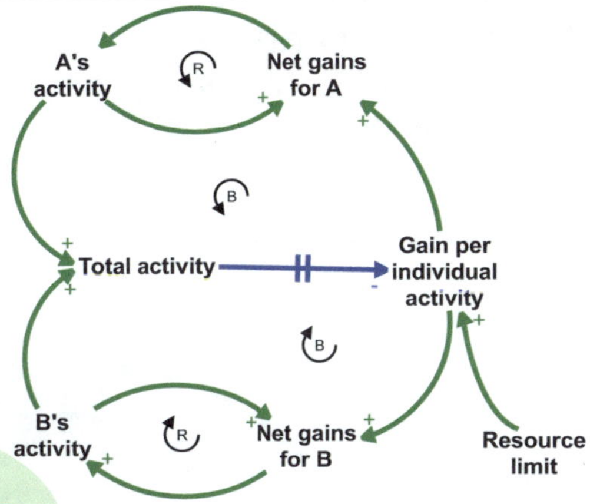

Title:__________________________________

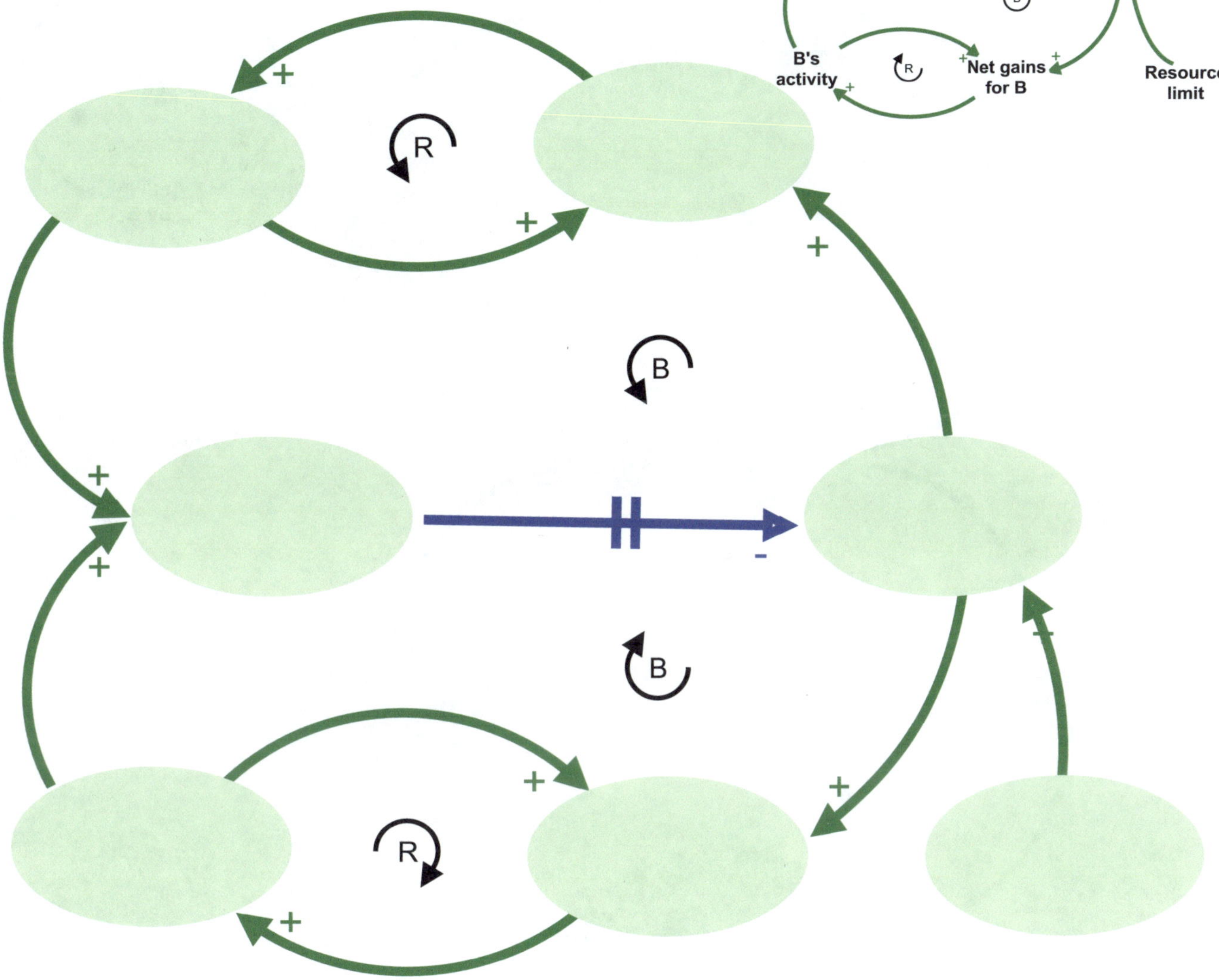

Story:

Template: Escalation

Use the guide at right as you complete your own example of this archetype below. Create a title and tell the "story" that explains your example.

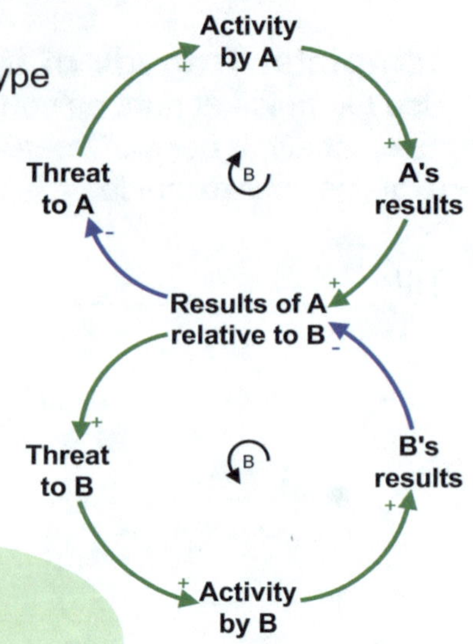

Title:________________________________

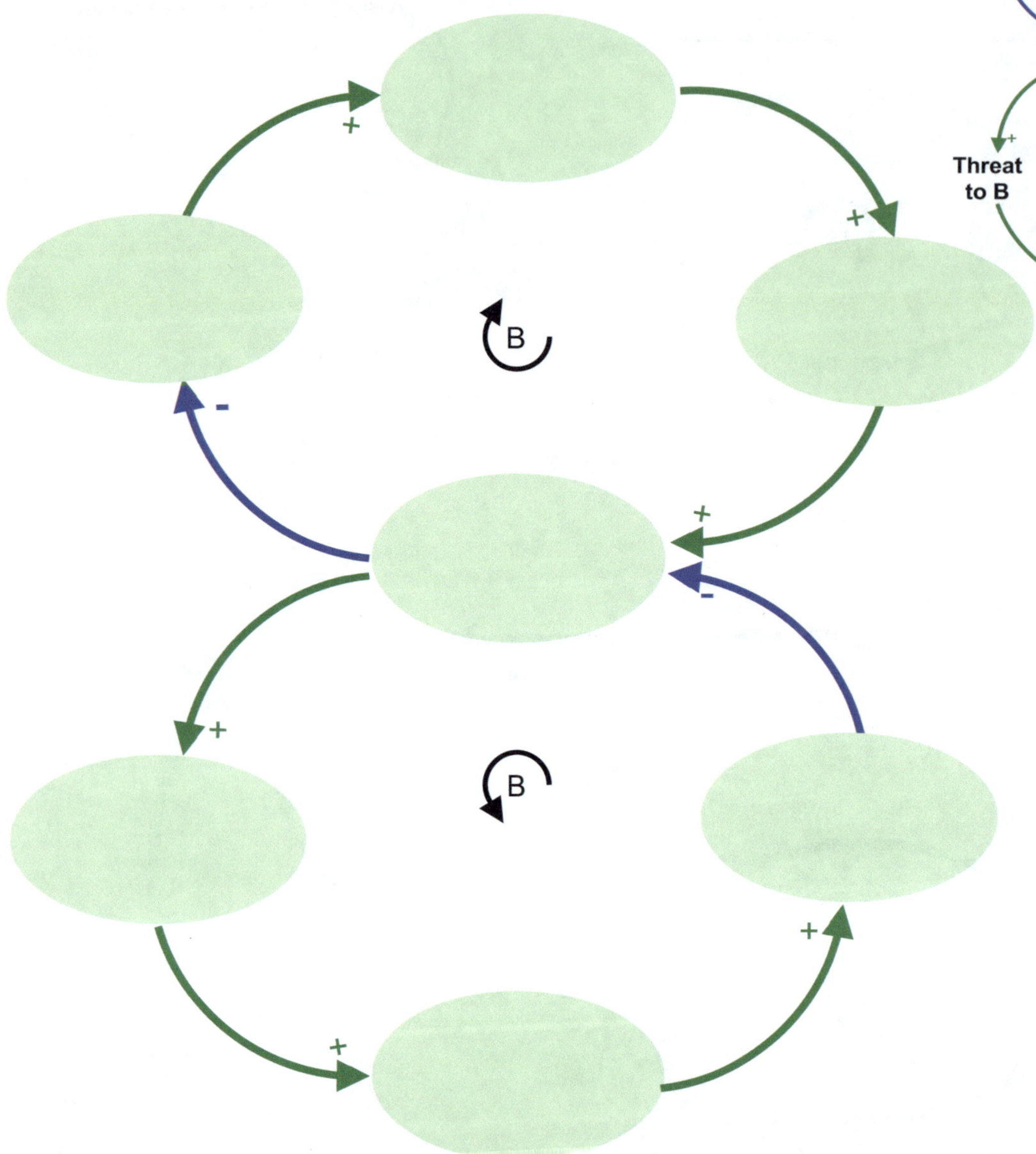

Story:

Template: Drifting Goals

Use the guide at right as you complete your own example of this archetype below. Create a title and tell the "story" that explains your example.

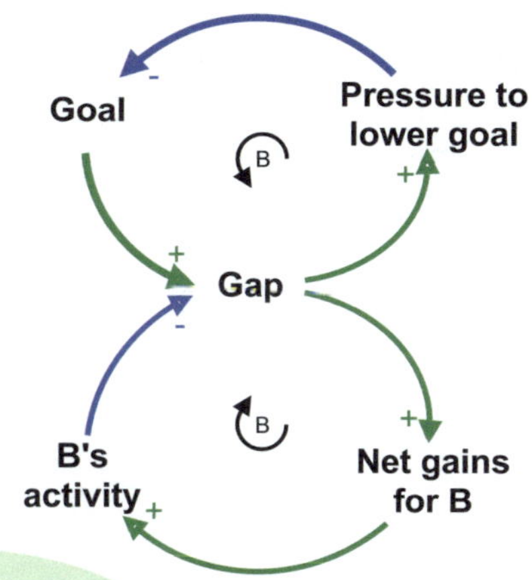

Title:______________________________

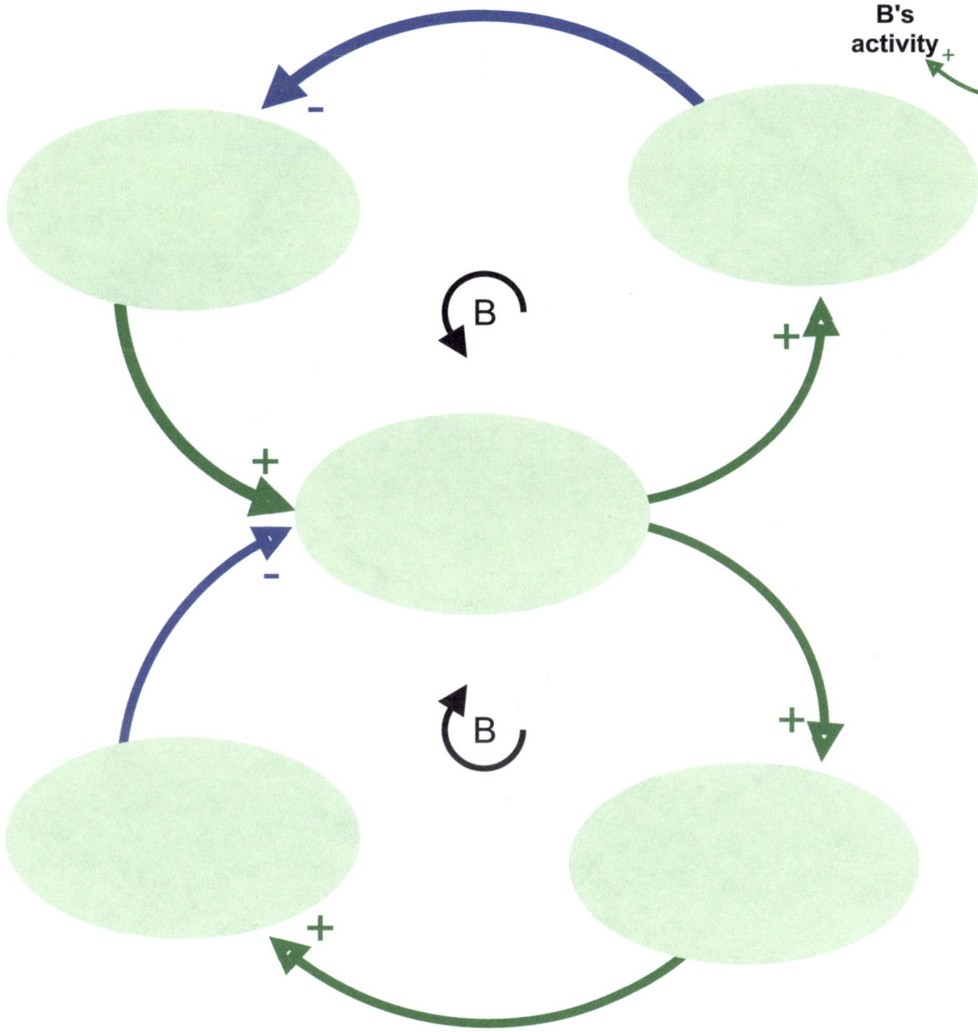

Story:

Template: Fixes That Fail

Use the guide at right as you complete your own example of this archetype below. Create a title and tell the "story" that explains your example.

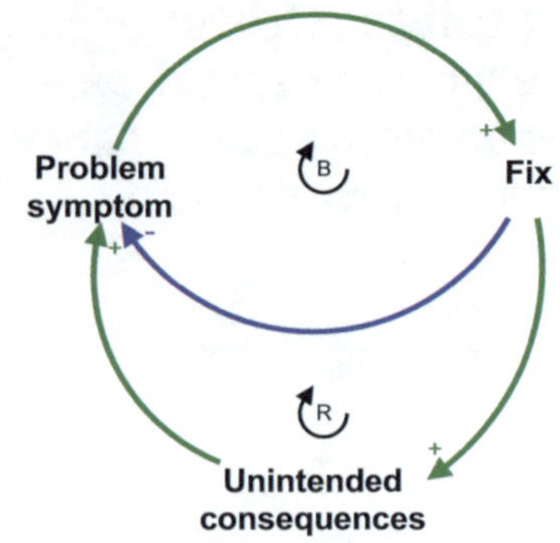

Title:______________________________

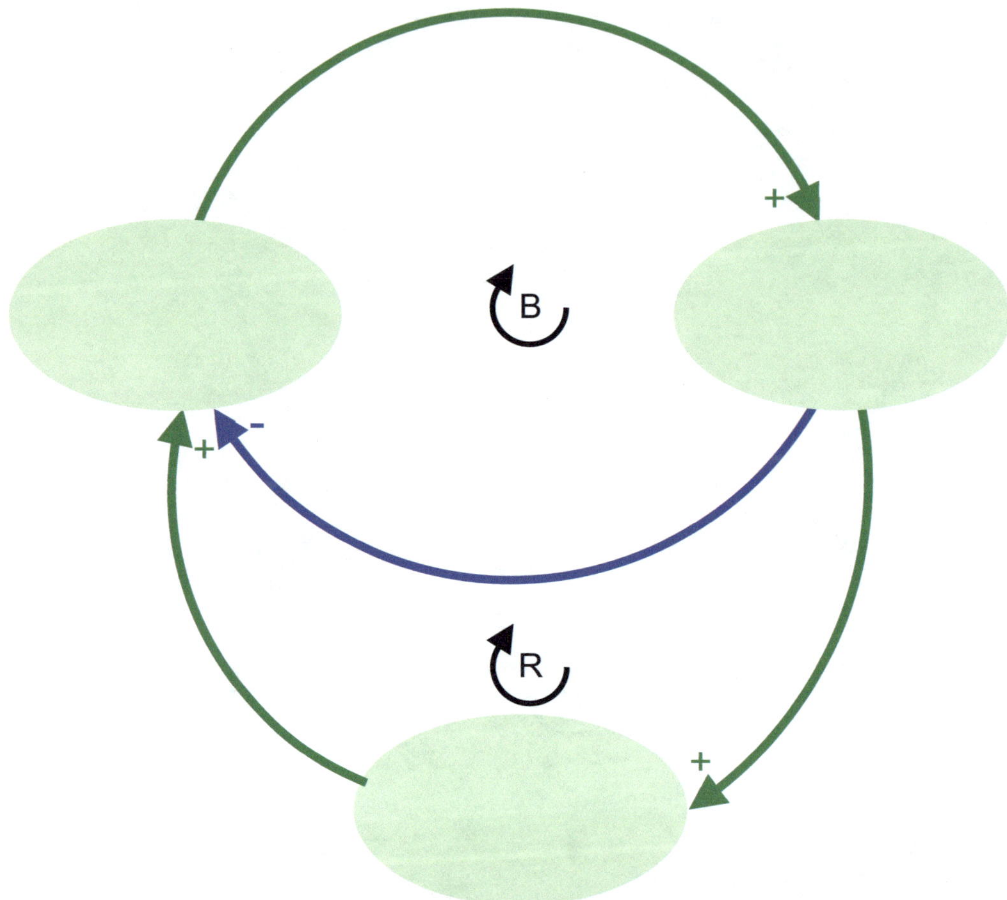

Story:

Template: Shifting the Burden

Use the guide at right as you complete your own example of this archetype below. Create a title and tell the "story" that explains your example.

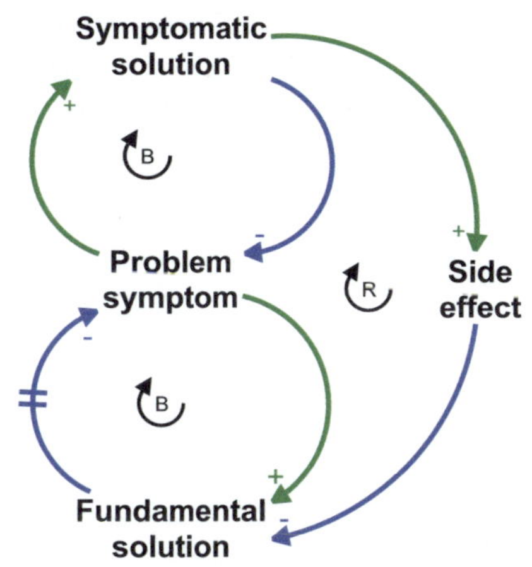

Title:______________________________

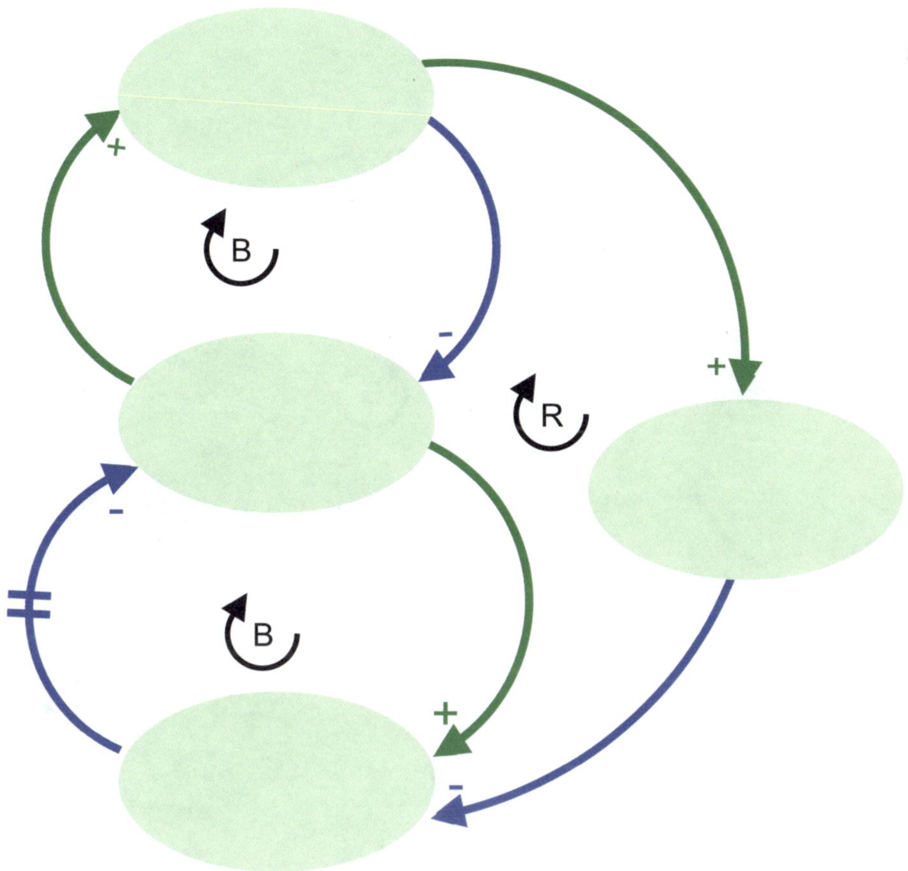

Story:

Template: Limits to Growth

Use the guide at right as you complete your own example of this archetype below. Create a title and tell the "story" that explains your example.

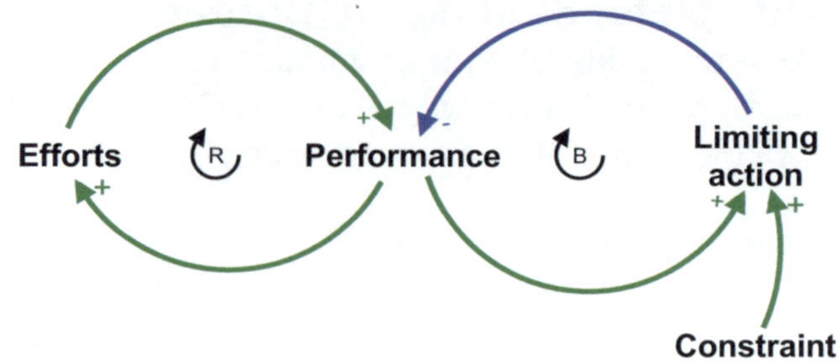

Title:________________________________

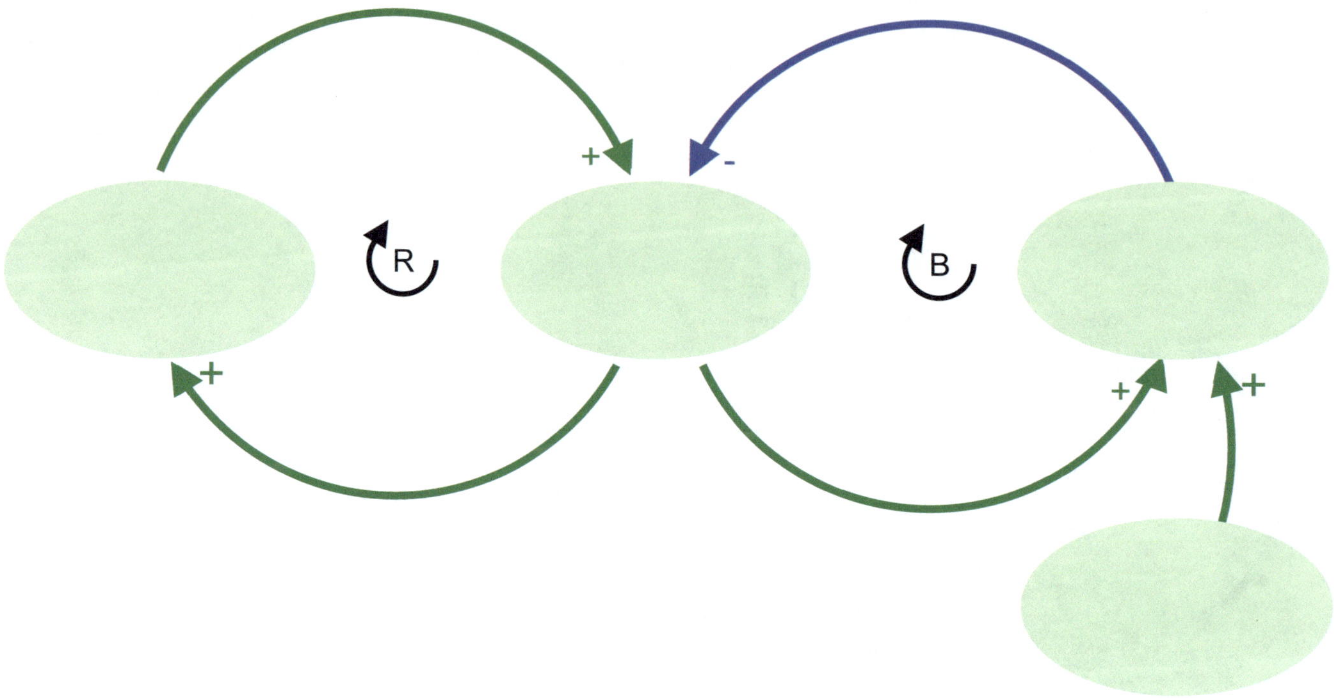

Story:

Template: Success to the Successful

Use the guide at right as you complete your own example of this archetype below. Create a title and tell the "story" that explains your example.

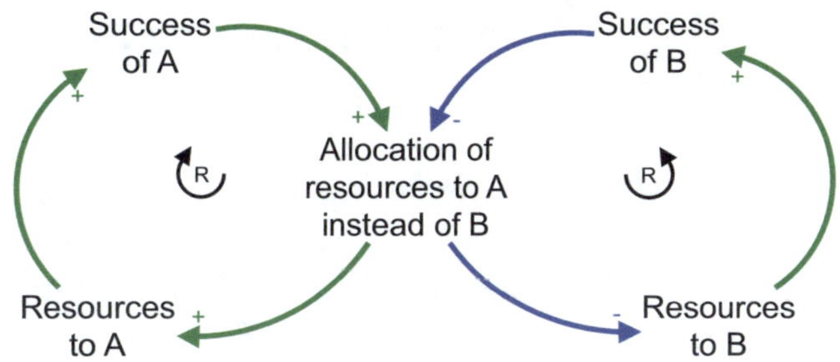

Title:_______________________________

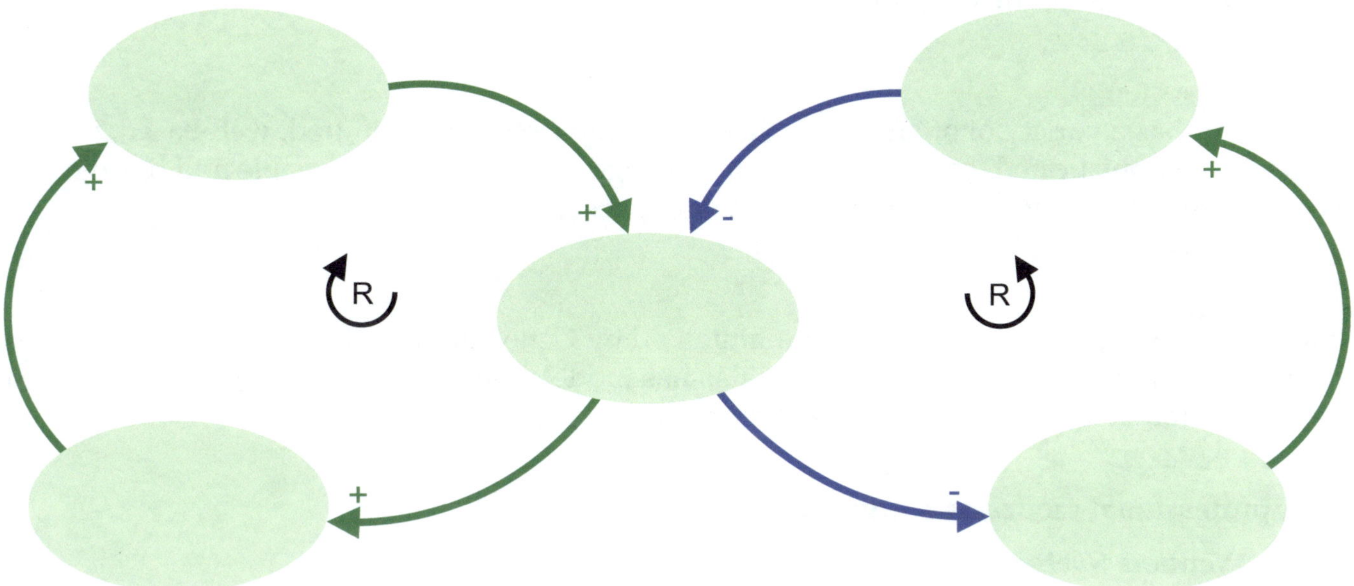

Story:

Introduction to System Dynamics Modeling Software

System Dynamics Modeling software comes in many shapes and sizes. One size does not fit all. These are a few that may be helpful for beginners who wish to get into more sophisticated modeling. The ones listed here are most familiar to the author. This in no way diminishes other available options.

Splash! and SplashExchange
These have been created by the Creative Learning Exchange and BTN, Ltd. to be visual, fun and interactive. Splash! is a free app useable on tablets and phones. It is terrific for creating very simple models and helping teach/learn the basics of stocks and flows as well as feedback. SplashExchange is a full mathematical modeling software with visual appeal. It will run in a browser on laptops and desktop machines.
clexchange.org/splash.

Stella (isee systems)
isee systems has a variety of offerings for Stella, ranging from a free, limited, web-based, modeling package to a sophisticated, complete software package. Stella is very accessible and has the easiest learning curve of the professional software packages.
iseesystems.com

Insightmaker
This free modeling software is built and maintained by Gene Bellinger. It comes with free tutorials, and many find it easy to grasp. It also has models that are primarily agent-based (each individual modeled), as well as those that are typical system dynamics models.
insightmaker.com

Other professional modeling software:

Vensim (Ventana Systems)
Vensim is a very sophisticated system dynamics modeling package that has the advantage of a free downloadable version (Vensim PLE) with many capabilities. They also have professional versions for the modelers who are creating complex and extensive models.
vensim.com.

PowerSim Software
powersim.com

Silico Software
strategydynamics.com

Note that the stock/flow and causal loop diagrams in this book were created using Stella software.

Appendix D: Instructional Resources

Overview

In addition to handouts and resources included with individual lessons and within Appendix C, these are provided as general instructional resources for use within any of the lessons. Teachers may choose to differentiate or modify them for their own purposes.

Assessment Strategies for the Lessons

Educators hold a significant degree of personal choice regarding methods for assessing student growth and achievement. High expectations motivate students to work hard, ask good questions, and learn with open minds. Trust the students. They will use the tools and resources in creative, unpredictable ways.

Assessment for the lessons in *Now What?* may be divided into three categories:

- Content
- Systems Thinking Tools
- Systems Principles

Bringing the Lesson Home and *Bringing the Chapter Home*, the debriefing sections at the end of each lesson and chapter, provide teachers with questions to gauge student understanding in each area. Here are suggestions for further assessment.

Content
Each lesson plan suggests ways for students to document and share their work or participate in an activity or simulation. Teachers can observe and collect students' projects and watch them in interactive simulations and presentations to measure their achievement and effort to understand key environmental and sustainability concepts.

Be sure to evaluate students' growth and skills in different areas in order to give effective feedback. For example, some students will have difficulty speaking to an audience if they are inexperienced, nervous, or shy. Find other aspects of their work to praise and reinforce. Remember that learning is an iterative process.

Keep the following areas in mind while assessing student work:

- Quality of research
- Analysis of data and information that supports conclusions
- Use of a variety of sources
- Integration of systems tools
- Clarity of presentation or product
- Ability to work with teammates
- Effective graphic content and images

Systems Thinking Tools
The standards in Appendix D can be used as a checklist and will provide specific feedback to students. Teachers can adapt the standards to use with students at different levels of mastery to create rubrics that are continuum or level based.

Encourage students to address and improve their graphs, diagrams, and models based on feedback received in a similar way to revisions expected in other academic work.

Systems Principles
Whole class and group conversations based on student projects are essential to surfacing the big ideas that are the key to understanding the complexity of environmental systems. Those discussions allow ideas to be shared and explored and provide teachers with a sense of the collective level of learning. Combined with the debriefing questions at the end of each lesson and chapter, student reaction to their peers' projects is essential to "bringing the lesson home."

When individual reaction is desired for assessment, teachers can gather written responses to the debriefing questions and others that have arisen. Encourage students to ask their own questions, a vitally important skill when dealing with complex systems. Finding solutions will not be simple and clear-cut, and student opinions will reflect that uncertainty.

Bringing the Assessment Home

Understanding complex environmental systems is difficult. Discovering and implementing sustainable and just solutions to environmental problems is even harder. Students need to investigate content topics rigorously, applying systems tools and principles as they do their research and learning. Teachers should use formal and informal assessment methods to provide feedback, keeping in mind that overall evaluation of student growth and performance should be broadly based.

- Integrate all three areas: Content, Systems Tools, and Systems Principles
- Emphasize interconnections
- Encourage students to see the big picture
- Reinforce positive and specific strategies

Upcoming generations will inherit a world that requires changes in the way humans interact with our environment. Encouraging systems citizens while they study in school gives them a head start toward accepting and meeting that challenge.

Strategies for Presenting Student Work

When teams or individual students have conclusions to share, finding creative structures for presenting their work will motivate them to produce clear, high quality writing and graphic content. Effective communication should be a goal for every project, both when delivered "live" and when posted or published for audiences.

The following list can serve as a starting point of options:

- **Deliver an illustrated lecture** - Have students explain their research to another class. An elementary grade audience would require them to simplify concepts and diagrams. How about an audience of adults who are not professional scientists?
- **Draw cartoons** – Editorial cartoons are very effective and will appeal to many students who enjoy drawing and exercising their sense of humor. Because the prior knowledge among your students will vary, you need to make a teaching decision about presenting examples before they create their cartoons.
- **Prepare a slide show** – Presentations may include cartoons and graphics.
- **Create and perform a song, rap, or poem**
- **Write an explanation in their own words, integrating a stock/ flow model or other systems diagram** – Students can write persuasive essays, structuring them as "op-ed" pieces, or letters to the editor of newspapers and periodicals.
- **Prepare and deliver a presentation to a town meeting** - Imagine your community needs to decide whether or not to use the resources of the local government to create and promote programs regarding environmental systems. Students can advocate for policies government should implement. For example, some cities and towns are moving toward renewable energy sources to power schools, municipal offices, and other public facilities. Explain why this is important.
- **Select an expert panel** – Choose students who have studied different aspects of a problem, including those with opposing conclusions. After brief statements, have them take questions from an audience of peers.
- **Build feedback into poster presentations** – Have students present their work on physical displays or digitally. Then encourage the audience of peers or guests to submit comments and questions after viewing the exhibits.
- **Publish a class magazine or newspaper** – Highlight student research and diagrams and disseminate online or print copies.

You can help other teachers by having your students share their projects. The CLE provides a website in which student examples are available for inspiration for other teachers. Consider having your students share work they have created (photograph and short description) to webmaster@clexchange.org with the subject line, "Now What: Student Projects."

Be sure you have students share according to school/district guidelines, indicating how the work should be credited, e.g., with name, school, or anonymously.

Handout: Strategies for Presenting Your Work

When you have conclusions to share, sometimes finding creative structures for presentations increases the impact of your ideas. Effective communication should be a goal for every project, both when delivered "live" and when posted or published for audiences.

Here are some options for sharing your work:

- **Deliver an illustrated lecture to a new audience** - Explain your research to another class. An elementary grade audience for example requires simpler concepts and diagrams. Or, address an audience of adults who are not professional scientists and use persuasive arguments.
- **Draw cartoons** – Editorial cartoons are entertaining and effective and can highlight your artistic ability and sense of humor.
- **Prepare a slide show** – Presentations may include cartoons among your graphics.
- **Create and perform a song, rap, or poem**
- **Write an explanation in your own words, integrating a stock/ flow model or other systems diagram** – Persuasive essays may take the form of "op-ed" pieces or letters to the editor of newspapers and periodicals.
- **Prepare and deliver a presentation to a real or simulated town meeting** - When your community needs to decide whether or not to use resources to create and promote programs regarding environmental systems, advocate for policies you believe government should implement. For example, some cities and towns are moving toward renewable energy sources to power schools, municipal offices, and other public facilities. Explain why this is important. If you don't have access to local government, simulate a meeting in your classroom.
- **Select an expert panel** – Team with other students who have studied different aspects of a problem, including those with opposing conclusions. After brief statements, take questions from an audience of peers.
- **Build feedback into poster presentations** – When you present your work on physical displays or digitally, encourage the audience to submit comments and questions.
- **Publish a class magazine or newspaper** – Highlight your research and diagrams and disseminate online or in print.

Consider sharing your final work (by sending a photograph and short description) to webmaster@clexchange.org with the subject line, "Now What: Student Projects."

Make sure to obtain permission from your teacher before sending and indicate how the work should be credited, e.g., with name, school, or anonymously.

Systems Thinking/System Dynamics Project Standards

See next page for system dynamics model standards

BOTGs

- The x-axis represents time.
- The y-axis represents the variable being investigated.
- The graph is a line graph.
- All variables are quantities that can increase or decrease with time.
- Scales are clearly labeled and related to the behavior. They can be numeric or descriptive.
- The line accurately shows the behavior described.
- The scale is long enough to show the pattern of behavior.
- When possible, the graph shows lines for more than one related variable.
- The graph focuses attention on the pattern of behavior rather than only on specific events.
- Variables are carefully chosen.
- The graph leads to questions about the dynamics of the system and possible causes of the behavior.
- The written project summary includes an explanation of what the graph shows.

Feedback Loops

- Each arrow shows a causal relationship between two variables.
- Does not use "more" or "less"
- Arrows show smooth causal links through the diagram with a minimum of crossing lines.
- All variables are nouns, quantities that can increase or decrease over time.
- Variables are carefully chosen and are significant.
- Variables combined when appropriate
- All polarities (+/- or s/o) are labeled.
- Connections are based on evidence, data, or reasonable hypotheses.
- The diagram includes feedback loops. Positive (reinforcing) and negative (balancing) feedback loops are labeled.
- Delays are indicated.
- A brief description tracing the feedback loops is included in the written project summary.

Stock/Flow Diagrams

- Significant stocks in the system are identified.
- Stocks are expressed as nouns, or noun phrases.
- Stocks are accumulations that go up and down over time.
- Stocks are changed only by flows.
- Flows have accurate titles.
- Converters show the relationships among elements of the model.
- Converters clearly show the influence behavior of the flows.
- The diagram shows causality (feedback) within the system.
- Positive and negative feedback loops are labeled.
- The diagram has smooth connections and limited crossing lines.
- The diagram provides insight into the possible causes of the behavior in question.
- The diagram is "operational"; graphs are generated which simulate behaviors.
- The written explanation of the structure of the system is clear and shows understanding and insight.

Iceberg Models

- *Events* are identified clearly.
- *Patterns* shows variables that change over time.
- Iceberg shows at least three levels.
- *Patterns* cited are significant and help to describe the problem.
- *Structures* are identified that cause patterns of behavior.
- *Structures* represented by feedback loops or S/F diagrams
- Counterintuitive and/or unintended outcomes are described.
- Delays between iceberg levels are noted.
- *Mental models* are linked to appropriate sections in the diagram.
- Explanation highlights possible leverage points that could change the behaviors.
- A short written description of the diagram and significant insights is included.

System Dynamics Models

Conceptualization

- The model has a clear purpose stated in behavioral terms.
- The question is about something that changes over time.
- The question involves feedback – the cause of the behavior is endogenous to the system.
- The question has a manageable focus – the boundaries of the system are clear and reasonable.
- System dynamics tools can provide useful insight into the problem.
- The problem is framed in a reference behavior mode – an initial behavior over time sketch of the behavior pattern in question. (See BOTGs)
- The initial mental model is briefly described. These are assumptions about how the system works and what is causing the behavior in question. Key elements of the system and their interrelationships are hypothesized.

Reflection throughout the Project:

- Is your mental model changing?
- Do your findings surprise you?
- Where are you encountering difficulty?
- Where can you get the help or information you need?
- Are things still missing from your model? How could it be improved?

Model Construction

Diagram(See S/F standards)

The Equations (The Assumptions)

- Equations mathematically describe causal relationships.
- All variables are named and represent real-world entities; they can be soft or hard.
- Units are listed for each variable and balance in each equation and throughout the model.
- All equations are simple and clear, briefly documented in simple terms.
- Complex equations are broken into simpler converters for clarity.
- Equations are arithmetic. Non-linear relationships are expressed as graphic functions.
- Graphic functions accurately capture complex non-linear relationships; curves are smooth.
- All parameters (constants) have reasonable real-world values.
- Switches are used judiciously.

Simulation Mechanics

- The solution interval (DT) is 1/3 or less of the smallest time constant to produce a smooth curve.
- The length of simulation is chosen to show the complete pattern of behavior.

The Structure

- The model starts with a simple structure and builds in complexity. The development sequence is shown if the model is complex.

Model Analysis

- The model accurately captures the behavior in question, usually requiring an iterative process of comparing the output to real-world values, identifying structures or parameters that need refinement, and simulating repeatedly.
- The behavior of all model variables is examined for reasonableness.
- The model is tested under extreme conditions.
- Leverage points are identified by determining which parameters have the greatest and least impact.
- Important causal loops are identified.
- Causes of the behavior are traced back through the loops.
- A causal loop diagram may be used to explain the feedback loops. (See CLDs.)
- Graphs are used to show and interpret the behavior under varying conditions.
- Supporting graphs show the behavior of all-important related variables.
- Scales are appropriate for ease of interpretation.

Completing the Project

- The presentation clearly communicates what was learned about the system. It is appropriate to the audience.
- All elements of the project are clearly and accurately labeled.
- Visual tools (graphs, pictures, tables, etc.) are used to support the presentation.
- The display is neat and aesthetically pleasing.
- The project is used to examine assumptions about different policies. The model is used to think about "what-ifs" and make decisions.
- If the project could not answer the question (which is OK too), there is an explanation of what was learned in the process.
- A brief written summary accompanies the display. It includes a description of the project and reflections.

Reflection on learning

- Did your project remind you of another similar system?
- How can you use what you have learned? Are there any leverage points to change the system?
- Can you recommend policy changes?

Sources

Cover Art

Hand holding Earth (foreground)
Gerd Altmann
Creative Commons 0
https://pixabay.com/photos/america-globe-hand-keep-present-2775323/

Trees and Clouds (background)
Adrian Pelletier
Creative Commons 0
https://pixnio.com/nature-landscapes/sky/blue-sky-white-clouds-clouds-treesCC0

Introduction

Reference to Tracing Connections
Joy Richmond, Lees Stuntz, Kathy Richmond, Joanne Egner, (editors)
Lebanon, NH and Acton, MA: isee systems and The Creative Learning Exchange, 2010: p. 51

K-12 ListServe
http://www.clexchange.org/cle/k12listserve.aspx

Chapter One: Air

East Asia Institute Roof
Immanuel Giel
Creative Commons 3.0 - https://creativecommons.org/licenses/by-sa/3.0/deed.en
https://commons.wikimedia.org/wiki/File:EastAsiaInstituteRoof.JPG

1.1 Cities

Lower Manhattan from Jersey City
King of Hearts
Wikimedia Commons CC-BY-SA-3.0 - https://creativecommons.org/licenses/by-sa/3.0/
https://commons.wikimedia.org/wiki/File:Lower_Manhattan_from_Jersey_City_November_2014_panorama_3.jpg
Image modified through cropping according to use permissions

1.2 Transportation

Red Line T inbound to Alewife at South Station in Boston Mass
Acela2038
Released into the public domain by author, Acela2038, at the Wikipedia project.
https://commons.wikimedia.org/wiki/File:Northbound_Red_Line_train_at_South_Station,_April_2005.jpg

1.3 Electrical Energy

Power plant
jplenio
https://pixabay.com/service/license/
https://pixabay.com/photos/power-plant-brown-coal-windr%C3%A4der-4287090/

1.4 Recycling

Municipal Recycling Facilities
EPA
Public domain
https://commons.wikimedia.org/wiki/File:Municipal_recycling_facilities,_Montgomery_County,_MD._2007,_Credit_USEPA_(14410405277).jpg

1.5 Environmental History

Hurricane Florence
Alexander Gerst
CC2.0 - https://creativecommons.org/licenses/by-sa/2.0/
https://flickr.com/photos/72482589@N07/30766053128

Figure 1: Graph of energy use
U. S. Energy Information Administration / Monthly Energy Review July 2021
Public domain
https://www.eia.gov/totalenergy/data/browser/index.php?tbl=T01.03#/?f=A&start=1949&end=2020&charted=1-2-3-5-12

1.6 Population

Crowd
Pablo Costa Tirado
Creative Commons 3.0 - https://creativecommons.org/licenses/by-sa/3.0/deed.en
https://commons.wikimedia.org/wiki/File:Crowd_is_equal_to_Times_Square_-_panoramio.jpg

Chapter Two - Land

Mountain vista
Public domain
https://commons.wikimedia.org/wiki/File:Moraine_Lake_17092005.jpg

2.1 Take Care of the Land

Clear Cut Forest
Public domain
https://commons.wikimedia.org/wiki/File:Clear_cut_terrain,_Cascade_Mountains,_1925_(TRANSPORT_1576).jpg

Book Review
Beth Macy
https://www.nytimes.com/2018/03/09/books/review/behemoth-joshua-freeman.html

2.2 The Trajedy of the Commons

Mountain vista
Public domain
https://commons.wikimedia.org/wiki/File:Moraine_Lake_17092005.jpg

Reference to paper, "The Tragedy of the Commons"
Garrett Hardin
https://www.sciencemag.org/content/162/3859/1243.full

"The Tragedy of the Commons on Georges Bank, and Elsewhere"
Donella Meadows
http://www.donellameadows.org/archives/the-tragedy-of-the-commons-on-georges-bank-and-elsewhere/

2.3 Protecting the Land During the Anthropocene Epoch

Traveler Footprint
Surath Lakpura Fernando
Creative Commons Attribution-Share Alike 4.0 International license
https://commons.wikimedia.org/wiki/File:Sigiriya_from_Pidurangala_with_Traveler_Stepping.jpg

Sidebar reference,
Marcia Bjornerud,
Timefulness. Pp. 128-131

2.4 Cities and Urban Planning

Train layout
Jim Henderson
Public domain
https://commons.wikimedia.org/wiki/File:Lionel_Xmas_shop_O_layout_1095_AoA_jeh.jpg

2.5 Studying Rapa Nui

Rapa Nui
Vanessa.blasco
Creative Commons Attribution-Share Alike 3.0 Unported license.
https://commons.wikimedia.org/wiki/File:Ahu_Tongariki,_15_moais.JPG

Sidebar reference
John Sterman
"A Skeptic's Guide to Computer Models." 1988
http://web.mit.edu/jsterman/www/Skeptic's_Guide.pdf

Chapter Three - Air

Cloudscape
NASA
Public domain
https://commons.wikimedia.org/wiki/File:Cloudscape_Over_the_Philippine_Sea.jpg

3.1 Protecting the Atmosphere

Clean air buttons
Public domain
https://commons.wikimedia.org/wiki/File:Clean_air_buttons_close.JPG

3.2 Climate Change

Wildfire in the Pacific Northwest
Bureau of Land Management
Public domain
https://commons.wikimedia.org/wiki/File:Wildfire_in_the_Pacific_Northwest_(8776185274).jpg

3.3 Bathtub Thinking

Shibuya Station
Dick Thomas Johnson from Tokyo
Creative Commons Attribution 2.0 Generic license.
https://commons.wikimedia.org/wiki/File:Shibuya_Station_Access_to_Tokyu_Department_Store_202003.jpg

Department Store Task
"Classic" Department Store Task (Sterman, 2002; Cronin, Gonzalez, and Sterman, 2009).
http://jsterman.scripts.mit.edu/~jsterman/docs/Sterman-2009-DoesFormalSystemDynamics.pdf

Also see related papers:

Sterman, John D., *All models are wrong: reflections on becoming a systems scientist*. System Dynamics Review Volume 18 Number 4 Winter 2002

Cronin, Matthew A., Gonzalez, Cleotilde, John D. Sterman. *Why don't well-educated adults understand accumulation? A challenge to researchers, educators, and citizens*. Organizational Behavior and Human Decision Processes 108 (2009) 116–130

3.4 - The Carbon Cycle

Greenhouse effect
Orvalrochefort on Flickr
Creative Commons 2.0 - https://creativecommons.org/licenses/by/2.0/
https://www.flickr.com/photos/orvalrochefort/3147880747/

Sidebar reference
Marcia Bjornerud
Timefulness. Pp. 142-143.

3.5 - Greenhouse Effect

Arid desert
stevepb (pixabay.com)
https://www.needpix.com/photo/download/848613/drought-dry-arid-desert-climate-environment-warming-water-shortage-global-warming

3.6 Global Warming

Temperature Graph
NASA
Public domain
https://earthobservatory.nasa.gov/ContentFeature/GlobalWarming/images/giss_temperature.png

Chapter Four - Sea

Intro and 4.1 Protecting Water

Wave
Public domain
https://commons.wikimedia.org/wiki/File:Giant_ocean_wave.jpg

4.2 Mosquito Nets and Overfishing
Mosquito net fishing
Ton Rulkens from Mozambique
Creative Commons Attribution-Share Alike 2.0 Generic license
https://commons.wikimedia.org/wiki/File:Mosquito_net_fishing_(8307117696).jpg

4.3 Sea Level Change
Glacier
Wikimedia Commons, public domain
https://upload.wikimedia.org/wikipedia/commons/5/5c/Perito_Moreno_Glacier_Patagonia_Argentina_Luca_Galuzzi_2005.JPG

4.4 Altering the Ocean
Coral Bleaching
NPS Climate Change Response
Public domain
https://commons.wikimedia.org/wiki/File:Coral_Bleaching_(14274360077).jpg

4.5 Running a Fishing Company
Fishing boat
Kritzolina
Creative Commons Attribution-Share Alike 4.0 International license
https://commons.wikimedia.org/wiki/File:Fishing_boat_off_the_coast_of_Naxos.jpg

Chapter 5 - Pulling it All Together
Earth from Apollo 17
NASA
Public domain
Retrieved 2016 from https://commons.wikimedia.org/wiki/File:The_Earth_seen_from_Apollo_17_with_transparent_background.png

5.1 Connecting Systems
Kids with "web"
BLM
Public domain
https://www.flickr.com/photos/blmutah/34395060511

5.2 Choosing a Topic
Art and Activism
by Fabrice Florin from Mill Valley, USA
Creative Commons Attribution-Share Alike 2.0 Generic license
https://commons.wikimedia.org/wiki/File:Art_and_Activism_for_Climate_Action_(30792597188).jpg

Epilogue
Quote
Andrew Deutz
Nature Conservancy Magazine, Fall 2021, p.62.

Appendices (and throughout)

Standards

Next Generation Science Standards (NGSS)
Overview
https://www.nextgenscience.org/get-to-know
Progression
https://www.nextgenscience.org/sites/default/files/resource/files/AppendixE-ProgressionswithinNGSS-061617.pdf
Crosscutting concepts
https://www.nextgenscience.org/sites/default/files/Appendix%20G%20-%20Crosscutting%20Concepts%20FINAL%20edited%204.10.13.pdf

National Science Education Standards
https://www.nap.edu/catalog/4962/national-science-education-standards

National Science Teachers Association
https://www.nsta.org/nstas-official-positions/environmental-education

Icons from Stella software
isee systems, inc.
Used with permission.

Icons

empathic
John Hain
Free for commercial use, no attribution needed
https://pixabay.com/illustrations/we-me-us-one-individual-separate-2078025/

waveform
https://openclipart.org/detail/260966/rgb-sound-wave-2

smileys
https://openclipart.org/detail/18499/glossy-smiley-set-3

arrow circle
https://openclipart.org/detail/236562/rgb-circle-arrows

knot
https://pixy.org/75955/
https://pixy.org/75951/
https://pixabay.com/vectors/trefoil-circular-knot-topology-40459/

shell
https://www.needpix.com/photo/174809/shell-sea-orange-pink-spiral-ocean-animal-coral-nautilus

head with question
GDJ
https://pixabay.com/vectors/human-head-man-male-cranium-1211467/
no attribution

slinky
Jake Ranallo from Noun Project
Creative Commons
https://thenounproject.com/term/slinky/48148/

Brain
Creative Commons
https://www.nicepng.com/downpng/u2q8y3w7q8y3a9y3_brain-clipart-neurology-transparent-background-clipart-brain/

simulation icon modified from original on Freepik
Original technology vector created by macrovector
https://www.freepik.com/free-vector/virtual-reality-visualization-simulation-icon-set_4329612.htm

Greenhouse Image (with Project)
NASA
Public domain
https://sealevel.jpl.nasa.gov/files/archive/activities/ts1hiac1.pdf

Barry Richmond's Systems Thinking Skills
Adapted from Tracing Connections by Creative Learning Exchange, 2021

The following was provided by the Creative Learning Exchange Used with permission.

- Predator/Prey Simulation screen and DynamiQueST Photo
- Systems Thinking/System Dynamics Project Standards
- Tool and Skill badges
- Various Templates, as indicated

About Us

Alan Ticotsky

Alan Ticotsky has worked as an educator since 1975. As a classroom teacher, he taught in grades 1 – 6. He also served as a K-8 Curriculum Coordinator and most recently as a systems thinking mentor for grades 5 - 12. During these years, Alan became a strong advocate for hands-on, inquiry-based activities and created a repertoire of classroom techniques designed to promote a climate of collaboration and investigation.

Since the 1990s, Alan has employed systems thinking and system dynamics tools with students and in adult workshops. Systems thinking is a holistic approach, a process of considering how things affect each other. The field of system dynamics studies how things change over time and the effect of feedback loops within a system. Tools include graphing, diagrams, and computer models. He is coauthor, with Rob Quaden, of *The Shape of Change*, a book that engages students in games and simulations that teach basic system dynamics concepts and encourages students to transfer this learning across the curriculum. Alan has also written four science activity and instructional books - *Science Giants*, a trilogy of science books on physical science, life science, Earth and space and *Who Says You Can't Teach Science*.

Alan has led workshops, designed curriculum, and been a consultant for many organizations, including multiple public schools, NASA, Boeing, New England League of Middle Schools, and the National Science Teachers Association.

Alan lives with his wife in Massachusetts near their grown children and grandchildren.

The Creative Learning Exchange

CREATIVE LEARNING Exchange
Making Thinking Visible

The Creative Learning Exchange (CLE) was founded as a non-profit organization in 1991 to encourage an active, learner-centered process of discovery for 5–19 year old students that engages in meaningful, real-world problem solving through the mastery of systems thinking and system dynamics modeling. Since its inception, the CLE has worked to encourage teachers and educators to use systems thinking and system dynamics in classrooms and schools throughout the United States as well as internationally. The CLE has done this through its website that offers free curriculum, its products that include books and games that promote systems thinking, and a biennial conference to help educators and students learn and utilize systems thinking and system dynamics in the classroom and the school organization.

Lees Stuntz, CLE Director, supported this work through feedback and edits to the manuscript and publishing the final work.

Anne LaVigne, Curriculum Designer, assisted with general feedback, edits, layout, and design.

Bunny Lawton added her eagle eye to final editing.

Other Publications by The Creative Learning Exchange

These are just some of the additional resources available from The Creative Learning Exchange to explore how systems work and possible ways to improve them. To see more, visit the CLE website at http://www.clexchange.org/cleproducts/.

Critical Thinking Using Systems Thinking and Dynamic Modeling

Communicating Critical Thinking: Visual Tools for Student Projects

Model Mysteries

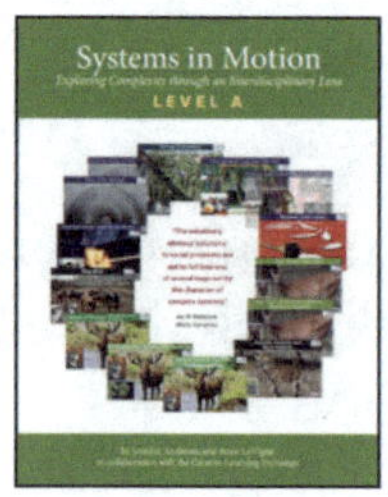

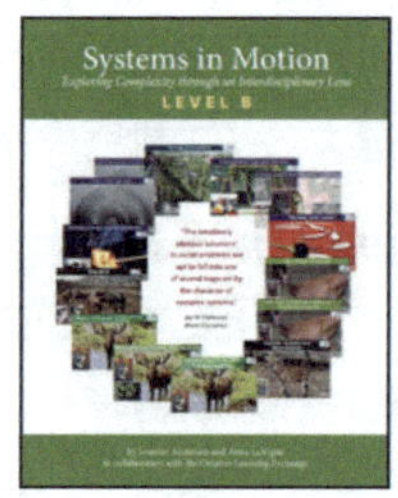

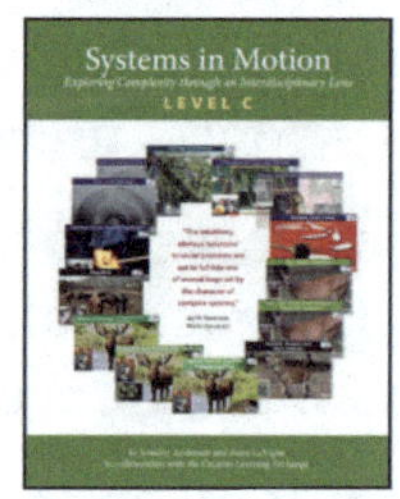

Systems in Motion - Level A, Level B and Level C

The Shape of Change

Dollars and Sense I and II

To explore additional resources from other organizations, visit the CLE website at http://www.clexchange.org/resources/additionalresources.asp#

Made in United States
Cleveland, OH
27 September 2026

44836273R00144